Brazil and the World System

Critical Reflections on Latin America Series
Institute of Latin American Studies
University of Texas at Austin

BRAZIL and the WORLD SYSTEM

Edited and with an introduction by **Richard Graham**
With chapters by
**Fernando A. Novais, John R. Hall,
and Luís Carlos Soares**

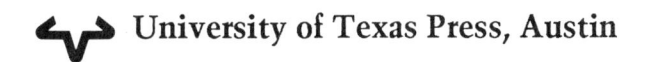

University of Texas Press, Austin

Copyright © 1991 by the University of Texas Press
All rights reserved
Printed in the United States of America

First Edition, 1991

Requests for permission to reproduce material from this work should be sent to:

Permissions
University of Texas Press
P.O. Box 7819
Austin, Texas 78713-7819

♾ The paper used in this publication meets the minimum requirements of American National Standard for Information Sciences—Permanence of Paper for Printed Library Materials, ANSI Z39.48-1984.

Library of Congress Cataloging-in-Publication Data

Brazil and the world system / edited and with an introduction by Richard
  Graham ; with chapters by Fernando A. Novais, John R. Hall, and Luís
  Carlos Soares.
      p.   cm. — (Critical reflections on Latin America series)
  Includes bibliographical references and index.
  ISBN 0-292-70785-1
  1. Brazil—Economic conditions. 2. Slavery—Brazil—History.
3. Brazil—History—To 1821. I. Graham, Richard, 1934–   .
II. Novais, Fernando A. III. Hall, John R. IV. Soares, Luís Carlos.
V. Series.
HC187.B8658 1991
330.981—dc20   91-20366
                                                              CIP

Contents

Preface	vii
Contributors	ix
Introduction *Richard Graham*	1
1. Brazil in the Old Colonial System *Fernando A. Novais*	11
2. The Patrimonial Dynamic in Colonial Brazil *John R. Hall*	57
3. From Slavery to Dependence: A Historiographical Perspective *Luís Carlos Soares*	89
Bibliography	109
Index	119

Preface

When Fernando Novais occupied the position of Tinker Visiting Professor of Latin American Studies at the University of Texas at Austin, it seemed appropriate to invite a number of other scholars to speak on campus and to examine the concept of a world system from the particular perspective of Brazilian history. Among these speakers were José Jobson de Arruda, John R. Hall, Frédéric Mauro, and Laura de Mello e Souza; Luís Carlos Soares was unable to visit our campus but sent us a paper that was presented on his behalf. We were fortunate to be able to draw on Enrique Semo, who was in Austin on a MacArthur Fellowship, to act as commentator. Out of that rich intellectual interchange grew the idea of a small book that would present some of its salient points to an English-reading audience. The present book attempts to do just that.

For making the original sessions possible, I wish to thank Prof. Gregory Urban, chairman of the Brazilian Studies Committee of the Institute of Latin American Studies, who enthusiastically backed the idea, Fernando Novais who did far more than could be properly asked of a visiting professor, and the staff of the Institute of Latin American Studies who made the local arrangements and have seen this effort through to publication.

—R.G.

Contributors

Editor Richard Graham is Frances Higginbotham Nalle Centennial Professor of History at the University of Texas at Austin. He received his BA from the College of Wooster and his MA and PhD from the University of Texas. His publications include *Britain and the Onset of Modernization in Brazil, 1850–1914, Politics and Patronage in Nineteenth-Century Brazil, New Approaches to Latin American History,* and *The Idea of Race in Latin America, 1870–1940.*

Fernando A. Novais is professor emeritus of history at the Universidade de São Paulo. His first article, published in the *Jornal de Filosofia* (1957), dealt with "Colonization and Economic Development," presaging his continued interest in this topic. He has published in the *Staden Jahrbuch*, the *Revista de História* (São Paulo), and the *Revista do Instituto de Estudos Brasileiros*, among other journals. His best known book is *Portugal e Brasil na crise do Antigo Sistema Colonial* (1979). His latest one (coauthored with Carlos Guilherme Mota) is *A independência política do Brasil* (1986).

John R. Hall is professor of sociology at the University of California, Davis. He received his BA from Yale and his MA and PhD from the University of Washington, and has also taught at the University of Missouri, Columbia. He has published widely in such journals as the *American Sociological Review, Sociological Analysis, Human Studies, Society, Latin American Research Review, Historical Methods, History and Theory,* and *Cultural Hermeneutics*. He is also the author of *The Ways Out: Utopian Communal Groups in an Age of Babylon* (1978).

Luís Carlos Soares is associate professor at the Universidade Federal Fluminense, where he received his BA and MA degrees. He received his PhD from the University of London. He is the author of *Rameiras, ilhoas, cocotes, polacas, e bagaxas: A prostituição no Rio de Janeiro do século XIX* (São Paulo, in press).

Brazil and the World System

Introduction

Richard Graham

The Brazilian past has been marked by three constants: exports, slavery, and patriarchy. Economic historians have typically emphasized Brazil's reliance on the export of raw materials and foodstuffs and the consequent reliance, until the present century, on the importation of manufactured goods. Social historians have long noted the presence of bondsmen, the relationship of slaves to masters, and the contours of Brazilian race relations. And political historians have often examined the tension between central government (whether in Portugal or later in Rio de Janeiro) and local leaders heading powerful clans and households. The present book brings these various themes together, but does so in a new way, reflecting critically on the proposition that Brazil's history is to be explained by its insertion in a single world system.

Portugal was the first European power to establish a plantation colony overseas. Whereas the Portuguese discovery of Brazil in 1500 was merely a serendipitous event in that nation's effort to establish trading "factories" in the Far East, Portuguese adventurers soon began shipping brazilwood from the American continent to supply dyes for the expanding textile industry of southern Europe. Meanwhile, since the fifteenth century, settlers on the Madeira Islands had been producing sugar to satisfy a large European demand for sweeteners, and it did not take long for similar undertakings to appear in Brazil: One of the first lord proprietors to be granted authority in Brazil in 1532—Duarte Coelho—introduced seed cane, brought in experienced workers from Madeira, built his first sugar mill, and undertook large-scale sugar production. Historians traditionally described the ensuing economic record as one marked by cycles of "booms" followed by "busts": After brazilwood and sugar (especially prosperous before 1660), came the gold and diamond rushes of the eighteenth century, and then the rapid expansion of profitable coffee exports from 1820 to 1930. This book discards this periodization and the implied "accidentalism" of such cycles to consider to what extent constant factors in the world economy shaped and defined Brazil's economic history. What accounts for its centuries-long reliance on exports, the slow development of its industry, and the continuing failure to develop an adequate supply of its own capital?

Even as they cut brazilwood, the Portuguese began to confront the problem of labor. From bartering with the Indians they soon moved to enslaving them. By 1538 they had turned to slaves from Africa, a trade that grew to horrendous proportions driven by the insatiable demand of sugar millers, mine owners, and coffee planters. Before the slave trade definitively ended in the early 1850s, some 3.6 million Africans had been forcibly transported to Brazil. In the late nineteenth century historians tended toward racist explanations for the course of Brazil's history, but, beginning with the work of Gilberto Freyre in 1930, they subsequently focused instead on the unequal power relationship between masters and slaves and the conditions of life in the slave quarters. But again, those are not the concerns here: The chapters in this book seek rather to find the connection between the presence of slavery in Brazil and the emergence of capitalism in Europe.

In establishing captaincies general in 1532, the Portuguese crown delegated authority and jurisdiction to others. Subsequently it tried to get that power back, and a continuing struggle between centripetal and centrifugal forces resulted, in some senses lasting until the present day. The conflict between Portuguese from Portugal and Portuguese born in Brazil began to emerge surprisingly early (certainly by the seventeenth century) and came to a head as an independence movement in the first third of the nineteenth century. Did this friction center on the class rivalry between Brazilian planters and Portuguese merchants, a rivalry in which colonial bureaucrats were caught up, or did it result from general nationalist feeling and the liberal ideology of the Enlightenment? The same question can be addressed to ensuing periods. The efforts of the Brazilian elite to determine the locus of power, whether in the national capital or in regional centers, led to a period of political instability that immediately followed independence in 1822, easing only as the prosperity of Rio de Janeiro coffee planters ensured the triumph and endurance of a centralized empire led by Pedro II from 1840 to 1889. But the ensuing "First Republic" (1889–1930) marked the reemergence of strong regional interests, curtailed again in 1930 with the ascension of sometime president, sometime dictator Getúlio Vargas. Again, it is not to the tension between centripetal and centrifugal forces that the authors of this book turn, but to the question of whether the exterior ties of Brazil should primarily be understood as sociological-cultural (as in a heritage of patrimonialism) or class based. Did the course of its political history depend on the rise of capitalism in Europe and finally within Brazil itself or derive from an entirely diverse dynamic? In short, chapters in this book take divergent and contrasting views on the role of the state, whether autonomous and controlling or the creature of a dominant economic class.

Many scholars have disagreed with the argument Fernando Novais makes in chapter 1 on the nature of the colonial system, Brazil's place within it, and the origins of slavery, but no one studying the economic and social history of that country can afford to ignore his views. Since its first publication in 1974,[1] heated debates have taken place in major public forums in Brazil and elsewhere. The reason for their intensity is not hard to find. In his incisive survey of European history from the late Middle Ages to the Industrial Revolution, Novais forcefully makes the case that the relationships between the emerging capitalist economy and the overseas colonial world were intimate and reciprocally causal. He then argues that slavery or other forms of compulsory labor in the colonies flowed directly from that relationship. The argument is made through a series of logical steps relying on deft summaries of well-known facts, but facts interpreted in a new way. Based on an extensive bibliographic foundation, he finds his way through the centuries and across great distances with confident strides. Paying much more attention to nuance and precision than others who have posited a "capitalist" origin to Brazilian colonization, Novais precisely describes the nature of the colonial system and its internal mechanisms. This chapter has become the locus classicus for the argument that the origins of New World slavery lie in the rise of European capitalism.

Novais's basic premise is that an understanding of the colonial experience depends on grasping the larger meaning of colonization in general and exploring the basic mechanisms underlying metropolitan-colonial relations. For him, the significance of the particular derives from the general. And the basic function of the overall colonial system was the initial or primitive accumulation of capital in Europe. He sees the period stretching from the Renaissance to the Industrial Revolution as one of slow transition from feudalism to capitalism in Europe, a transition dominated by the interests of merchants. Here he sides with Maurice Dobb in his famous 1950s debate with Paul Sweezy regarding the nature of that transition and the role of commerce within it.[2] He also, like the other authors in this book, defines capitalism as a system that relies on wage labor and, therefore, carefully qualifies this transitional period as one of "commercial capitalism."

Novais incisively argues that the creation and functioning of the colonial system was an integral part of that larger transition taking place in Europe and that the actual historical process of forming capitalism depended on it. For the European economy at that time was incapable of transforming itself entirely on its own, requiring colonies as levers eventually to bring forth modern capitalism. Without that initial piling up of capital resources, capitalism could not have emerged when and as it did. And that accumulation required colonies. Mercantilism was the legal and theoretical confirmation of the system's need. Colonies meant trade, and colonial trade meant monopoly. The system always operated on the principle that trade to the colony was the exclusive privilege of the metropolis and often of the king himself, although he could delegate that privilege to others. Along with that monopoly went a coercive apparatus to enforce it, so a governmental structure was created that historians have often identified as colonialism; but that apparatus was merely the means to an end. Whether considering the Portuguese or the Spanish and French experience, Novais finds the same principles at work: a monopoly system was designed to foster the accumulation of capital in Europe. Even England not only established similar plantation colonies in the Caribbean but, through the Navigation Acts, demonstrated its obsession with the exclusivist principle. The North Atlantic English colonies to be sure produced goods for other colonies, but these colonies in turn produced for Europe, the difference being that in this case, the profits accumulated in the colony itself. Nor do contraband and smuggling contradict the logic of the monopoly system, but rather confirm it: What smugglers desired was to participate in the advantages of that monopoly and in the exorbitant profits made in colonial trade.

Finally, the way in which the exports were produced was also defined by the colonial system, for this system led inexorably to compulsory forms of work. The availability of land encouraged free laborers to acquire it themselves and to become independent producers, raising crops not primarily for European markets but for their own consumption (or demanding wages so high as to preclude the accumulaton of capital in the European economy). Slavery or something approximating it was the only way out for a truly colonial, that is, export-oriented, economy. Furthermore, it was the concentration of wealth among the slaveowners within the colony that allowed the system to function, for it was they who consumed expensive imported European goods. But this labor system hindered investment in technology (resulting in low productivity), used up natural resources, and eventually limited the further growth of markets for European goods. Thus, the whole rationale of the colonial system (the general expansion of the market economy) was in the end subverted by slavery even though slavery was a logical outcome of that economic impulse. Once the Industrial Revolution began—a revolution made possible by the accumulation in Europe of capital derived from profits in colonial trade—then, Novais concludes, a new search for markets began that went far beyond the capability of regimes based on compulsory labor. Thus, from the late eighteenth century the whole colonial system entered a long period of crisis.

In chapter 2, John R. Hall adopts an entirely different intellectual method. He particularly takes exception to the whole concept of a world system. Indeed, he eschews any universal theory of historical change or any holistic approach in which the totality defines the parts of a system, because, he argues, such an approach submerges key aspects of reality. He prefers to analyze the past without privileging any particular explanation in advance. Rather than a holistic approach, he prefers a "neo-Weberian" one: for Max Weber offered an "array of concepts about action, types of productive organization, types of economic exchange, classes, status groups, parties, political action and alliances, structures of [social] organization, [and cultural forms]" that do not go together in a package but can be combined and recombined in an endless list of actual historical situations.

In considering colonial Brazil, Hall argues for the relevance—not as a total explanation, but as a factor worth specific and prolonged attention—of issues of political power and cultural structures that are ignored within a world-system approach. Specifically, he urges us to focus on how the powerful have developed, preserved, and exploited patrimonial relationships at the expense of fostering the expansion of society's economic resources. Thus, Brazilian development has a dynamic that cannot be reduced to that of the world economy, for patrimonialism does not derive from the logic of any world system or from a drive toward accumulation on a world scale.

What is patrimonialism? Hall explains that whereas in a feudal system the authority of a king is lessened by the formal delegation of power to his vassals, and in a rational bureaucratic one the king exercises that authority through an impersonal hierarchy loyal to him directly, in a patrimonial arrangement authority is maintained in the hand of the sovereign by the constant purchase of allegiance from dependents through the temporary grant of sources of wealth, that is, by the distribution of income opportunities that can also be withdrawn. Rather than seeing the economic activities of the fifteenth- and sixteenth-century kings of Portugal as aiming toward the accumulation of capital, Hall sees them as a method of political control. The ultimate royal monopoly of the avenues to wealth (especially through the regulation of trade) was part and parcel of the patrimonial system, a system of values and a pattern of action that would then dominate in Brazil to the present day. Patrimonialism, unlike feudalism, encourages trade as long as it is controlled by the crown. Sometimes kings favored Brazilian planters over Portuguese merchants, sometimes not, depending always on crown interests and the calculus of power. The crown co-opted the merchants rather than being controlled by them as Novais implies. Indeed, Portuguese merchants always remained weak in relationship to the king.

For Hall, therefore, one of the central issues in colonial history is how local interests—merchants and planters in Brazil—were subordinated to the king's authority. Hall argues that the state monopolized key resources essential to the prosperity of those classes, and encouraged local elites in turn to monopolize economic resources rather than distribute them among other groups. Here lies the reason for the very slow emergence of free markets in labor, commodities, or land. Because of patrimonial arrangements, Brazilian economic elites could respond only slowly to world market conditions. Noneconomic controls and politically arbitrary decisions structured economic life, thus inhibiting the rise of capitalism. The emergence of a class society in Brazil derived not from the world economic system but from the internal dynamics of an elite caught up and actively participating in a political system devoted to the use of wealth for the sake of concentrating power. They did not struggle against the king but sought to secure positions and influence within that patrimonial structure centered on the crown. Through marriages, in fact, local elites merged with crown officials. Their action fits with the logic of status-group competition among leading families. Class does not explain patron-client relations even when these are congruent with class interests; rather, it is patrimonial power that shapes class opportunities.

Finally, the emergence of the capitalist world economy did not alter the basically patrimonial nature of the relationship between the state and economic interests. In the eighteenth century, with the rule of the marquis of Pombal, "state patrimonial capitalism" became more modern and rational without ceasing to be patrimonial. For that matter, the state's active role in diversifying the economy and fostering its growth within a patrimonial framework has continued from then to the present day and can be seen at work despite Brazil's present industrial strength.

In chapter 3, Luís Carlos Soares, writing from a contrastingly and sometimes dogmatic Marxist point of view, surveys the Brazilian historical literature on the slave mode of production and dependency theory, critically examines the work of other Marxists as well as traditional historians, particularly with regard to the idea of a Brazilian nation, and ends up proposing a variegated approach to the history of colonial Brazil. His chapter is especially useful in setting forth the alternative points of view so far adopted by the current generation of Brazilian historians. Once one understands the position from which he writes—which he makes quite clear—the linkages he establishes are provocative and insightful.

Introduction

Soares begins by analyzing the work of Ciro F. S. Cardoso and Jacob Gorender, especially as these authors considered the slave mode of production in the Americas. Ciro Cardoso distanced himself from Eugene D. Genovese's proposition that American slave modes of production represented a regression to archaic arrangements long abandoned in Europe. Cardoso argued for the existence of particularly *colonial* modes of production, understood as modes that derived from their structural dependency on European metropolises. Cardoso, Soares points out, took great pains to differentiate the slave mode of production in the Americas from that of the Greco-Roman world. At the same time, this mode of production was clearly distinct from capitalism. Gorender also focused on the colonial slave mode of production, seeking to elaborate the laws that would operate for it much as Marx had done for the capitalistic mode of production. He too drew a sharp distinction between the American system and that of the ancient world. Soares criticizes both these authors for failing to note that the ruling class in the colonies was not the dominant class there but the commercial bourgeoisie and nobility in Europe. At this point, Soares adopts the scheme proposed by Novais for understanding the overall colonial system. But more crucially, Soares challenges the formulations of Cardoso and Gorender for relying on the "ambiguous" concept of dependency.

Dependency theory is especially linked to the name of another Cardoso, Fernando Henrique Cardoso, although Soares also considers the work of Theotonio dos Santos and Ruy Mauro Maurini. Soares traces the emergence of this theory in the debates about development, modernization, and import substituting industrialization of the 1950s and 1960s. F. H. Cardoso proposed the concept of "dependent capitalism," presumably—implies Soares—as a particular variation of the capitalist mode of production. But Soares, quoting Francisco Weffort, especially takes issue with the notion that the subunits of analysis are nations or countries and not classes.

Returning to Ciro Cardoso's and Gorender's notion of colonial and dependent modes of production, Soares argues that they forgot that these modes were being constructed in the Americas when a capitalist mode of production did not yet exist in Europe, they ignored the question of class dominance in Europe, and they implied that there were dependent *nations* in the Americas even before political independence. In contrast, Soares proposes that the primacy of commerce during the transitional phase between feudalism and industrial capitalism is central to understanding the colonial reality. It was in relation to this primacy that various modes of production emerged in the Americas and specifically in Brazil. Especially important is to recognize that "Brazil" did not constitute a single economic and social formation in colonial times as it does today. Soares then examines the way in which the idea of Brazil as a nation aborning was read into the past by nineteenth-century nationalists and twentieth-century populists. Marxists uncritically adopted the same concept. But only in the twentieth century is it truly possible to speak of a world capitalist system and of Brazil as having a single mode of production—a capitalist one.

As is apparent from the points made in these chapters, the concept of a world system considerably predates the publication in 1974 of works by Immanuel Wallerstein.[3] Even as the UN Economic Commission for Latin America struggled in the immediate postwar period to find a key to the region's underdevelopment, it understood that the principles of comparative advantage worked against the interests of that region. Fernando Henrique Cardoso elaborated what came to be called the dependency theory in the 1960s,[4] and Andre Gunder Frank published his controversial study, *Capitalism and Underdevelopment in Latin America*, in 1967, arguing that Latin America had always been capitalist from its initial settlement by Europeans and firmly inserted in a capitalist world.[5] In 1970 Stanley and Barbara Stein picked up on this theme and published their useful summary on the colonial, that is, export-oriented, heritage of Latin America.[6] Soon Ernesto Laclau, however, took strong issue with Frank et al., arguing that they had misunderstood the nature of capitalism. Long-distance trade did not capitalism make, but wage labor.[7] Laclau in turn has been roundly attacked for seemingly continuing to see Latin America as somehow having a "feudal" past. And it was into this literature that Wallerstein's argument fit. He argued that beginning in 1492 a single economic system emerged that pulled together all regions of the world. He posited that there were always three layers of participation in that system: the core, the periphery, and the semiperiphery. Particular areas of the world could be found sometimes in one, sometimes in another level, but the relationships between the levels remained constant. Furthermore, there were specific methods of controlling labor appropriate to each level. The publication of his books has elicited much debate in the North Atlantic world, but needless to say they seemed a bit passé in Latin America.[8] Nevertheless, it seems appropriate to have used his phrase as part of the title for this book, for each of the essays here presented reflects critically on the concept of world system.

Notes

1. Its principal elements are to be found in a paper delivered in 1969: "Colonização e sistema colonial: discussão de conceitos e perspectiva histórica," in IV Simpósio Nacional dos Professores Universitários de História, *Anais* (São Paulo, 1969).

Introduction

2. Sweezy et al., *The Transition from Feudalism to Capitalism*.

3. See the convenient summary statement of his theory in Wallerstein, "The Rise and Future Demise of the World Capitalist System," 387–415. His book also appeared that year: *The Modern World-System: Capitalist Agriculture and the Origins of the European World-Economy in the Sixteenth Century*.

4. Stern, "Feudalism, Capitalism, and the World-System," 836.

5. Frank, *Capitalism and Underdevelopment in Latin America: Historical Studies of Chile and Brazil*.

6. Stein and Stein, *The Colonial Heritage of Latin America*.

7. Laclau, "Feudalism and Capitalism in Latin America," 19–38.

8. Stern, "Feudalism, Capitalism," 829–845.

1. Brazil in the Old Colonial System

Fernando A. Novais
Translated by Richard Graham and Hank Phillips

Colonization as a System

Relations between the metropolises and their respective colonies during a particular period in our history is best understood as constituting a system of interrelated forces. For the early modern period—between the Renaissance and the French Revolution—it seems useful, in the tradition of many historians, to refer to these relationships as the Old Colonial System of the mercantilist era. Even this first, purely descriptive approach already allows us immediately to establish an early distinction of no small importance. Not all colonization occurred within the boundaries of this colonial system; the more general phenomena, the enlargement of the area of human settlement on the globe through the occupation, population, and development of "new" regions—in the words of geographer Maximilien Sorre, "the organization of the ecumenical"—has given rise to a wide variety of distinct historical situations, and it is only to one of them that we refer here.[1] In the early modern era, however, the unfolding of the process elaborated a specific set of relations, and this mercantilist system of colonization is the key to understanding all European colonization between the first maritime discoveries and the Industrial Revolution. In understanding it we can discern certain essential common denominators that persist as fundamental constants within a complex variety of particular historical circumstances—circumstances that varied from metropolis to metropolis as well as among the colonies—during the sixteenth, seventeenth, and eighteenth centuries.

These colonial relations can be understood on two levels: first, through the extensive colonial legislation imposed by the various colonizing powers (Portugal, Spain, Holland, France, and England); and, second, in the actual practice of trade and governance, that is, in the commercial arrangements and politico-administrative ties that involved all parties. The basic aim of colonial legislation was to discipline these political and, above all, economic relations. Still, for what we are undertaking at this early point in our analysis (defining the workings of European colonization during the Old Regime), our overriding concern is with the legal norms of the times, for these were the embodiment of the objectives of the colonizing entities, expressing their aims in the colonizing process. Thus, the English Acts of Trade, the "laws that ban foreign ships... from Brazilian ports,"[2] Colbertian colonial legislation, charter company regulations, and so on, were all relevant examples of the immense body of European colonial laws that encompass the common denominators we seek here. Also, during the unfolding of the colonization process, the role and position of the colonies within the framework of European states' economic life was laid out theoretically by the pundits of mercantilist economics. The goals of colonial undertakings were thus specified, and the legislation of the times merely served as a vehicle for carrying into practice the principles formulated by mercantilist theory.

In order firmly to orient ourselves within the enormous framework of European colonial history, it seems useful, as we develop a tentative initial characterization, to begin with the typical model of the functioning and relations of the Colonial Pact as formulated by these theorists of mercantilist policy. One such theorist who formulated that policy with crystal clarity was [Malachy] Postlethwayt, who in 1747 wrote: "The colonies... should: first, provide the metropolis with a larger market for its products; second, provide employment for a larger number of its [the metropolis's] manufacturers, artisans and mariners; [and] third, furnish a greater quantity of needed articles."[3] To use a modern term, the colonies ought to constitute an essential factor in the development of the metropolis. So went the theory. To be sure, actual history runs along somewhat atypical and peculiar lines rather than along those contained within the framework of the models; and European colonization during the modern epoch offers a whole gamut of situations both approaching and differing from that scheme, varying in time and space and inexorably complicating reality. Yet to ignore that basic project that for several centuries gave form to European overseas policies, a project that forms thereby part of this same and complex reality, would mean not coming to know the deeper mechanisms inherent in the process and obtaining only a superficial view of each isolated event. Viewed as a whole and polarizing the European economies on one side and the peripheral colonies on the other, it cannot be denied that the colonial experience was fashioned according to that fundamental desideratum. Hence its relevance to this analysis.

It is important to emphasize at the outset, however, that mercantilist doctrine had as its immediate objective the formulation of economic policy. The development of theory had as its sole purpose justifying that agenda. These theorists did not proceed from grand explanatory theories logically developed in a deductive way, but worked almost in the opposite direction. By the same token these theorists concerned themselves with little outside the borders of their own nation; the move from "England's Treasure" (with which Thomas Mun concerned himself) to "The Wealth of Nations" (which would occupy Adam Smith) represented a broadening of intellectual horizons that betokened a different stage in the scientific formulation of economic theory, involving broader generalizations and expressly corresponding to distinct moments in the course of Western Europe's political and economic evolution.

Our interest here is merely to sketch the main outlines of the general doctrine in order to situate the role of mercantilist colonialism within it. Let us try to establish its essential features. The starting point, as we know, was the "metalist" idea—the identification of the level of wealth with the amount of gold and silver extant in each nation. Although the concept of wealth as identified with coinable metals was weighted disparately by the various thinkers who perfected the theory of mercantilism, the basic metalist idea oriented the political economy of each one. It involved a narrow conceptualization of the economic good and the supposition that profits are generated through the process of merchandise circulation, that is, that advantage necessarily results from losses suffered by one's trading partner. Not surprisingly, the mercantilist agenda tends directly toward maintaining a favorable balance—a favorable balance of contracts on the level of individual merchants and a favorable balance of trade on the level of international exchange. That was the means of promoting the entry of bullion that measured national wealth; hence, the adoption of a protectionist policy with primary emphasis on tariffs, and the stimulation of domestic production of those products most able to compete advantageously in the international market. Creating obstacles to the export of raw materials and boosting manufactured exports also followed; conversely, the policy fostered entry of raw materials and hindered or even banned imports of manufactures. The point of all of this was to lower the costs of internal production, with international competition as an end.

Mercantilism was not a political economy intent on social welfare. It aimed toward national development at all costs, legitimizing all manner of state stimuli and relying on state intervention to bring about lucrative conditions to enable enterprises to maximize the export of surpluses. This led to a policy fostering demographic expansion with an eye toward increasing the national labor force and impeding the rise of salaries.

Within this context one can easily view the position and significance of the colonies, their role being to constitute the metropoles' economic rear guard. While mercantilist policy, as practiced by the various modern states, took the form of unbridled competition with one another in Europe, there arose a need to reserve certain areas within which, by definition, mercantilist norms would be applicable. The colonies became guarantors of metropolitan self-sufficiency—which was the fundamental goal of mercantilist policy—thus permitting the colonizing state to compete advantageously with other powers.

The mercantilist notion, furthermore, did not emerge as an isolated element in the political and economic thought of theorists and statesmen. On the contrary, the belief was articulated in harmony with a body of ideas that developed and predominated in Europe during the period between the Discoveries and the Industrial Revolution. The colonizing project was tightly bound up, first of all, with the political principles of the absolutist era. The goal was clear. The powers' colonial politics sought to tailor colonial expansion to suit mercantilist aims; to cause relations between both poles of the system (metropolis-colony) to unfold in a manner consonant with the scheme. Even here, within this first description of the colonial system, we see that it reveals itself as a specific type of political relationship possessing two elements: a decision-making center (metropolis) and another area (the colony) that was subordinate. These political relations establish the institutional framework permitting the economic life of the metropolis to be fueled by colonial activities. If we are to plumb this long-term phenomenon more deeply, we will discover its connections with the very processes of modern colonization and other components of the totality. Such connections are not merely logically functional relations, but are constituted historically over time and in practice.

If overseas expansion and the colonization of the New World constitutes one of the most striking features of the sixteenth to the eighteenth centuries, one sees contemporaneously the predominance of absolutist systems in the political realm and, in the social sphere, the persistence of societies of "orders" founded upon juridical privileges. And within the universe of economic life, sometime between the gradual dissolution of feudal structures and the advent of capitalist production, there emerged (along with remnants of the former and elements peculiar to the latter) an intermediate stage that is commonly referred to as mercantile capitalism: For commercial capital, that is, capital generated most directly from the circulation of market goods, animated all economic life. The absolutist state, with its extreme centralization of royal power, which in certain ways unifies and disciplines a society organized into "orders," also executes a mercantilist policy encouraging the development of a market economy both internally and externally—externally by means of colonial development. Such were the parts of the whole that it behooves us to integrate. Truly, the relations between the unitary, centralized monarchy (or, earlier, between the *processes* of centralization and unification) and the adoption of mercantilist policies are clear. According to Eli F. Heckscher's formulation, mercantilism itself served as an instrument of that unification—albeit requiring a certain degree of national integration in order to emerge.[4] Their relations were, therefore, reversible, which leads us to believe that both emanate from a common source, namely, the critical process of transition out of the feudal structure. Similarly, overseas expansion allowed the breaching of those narrow limits that had constrained the movement of the mercantile economy up until the end of the Middle Ages.

It would be impractical, within the limits we've proposed, to attempt to include here an analysis of the crisis of feudalism. Let us simply state in passing, in agreement with the the analyses of Maurice Dobb, that on the whole, it arose not only from the commercial renaissance per se, but also from the manner in which the feudal structure reacted to the impact of the market economy.[5] The revival of commerce (i.e., the introduction of a strengthened mercantile sector into the economy and the development of an urban sector in the society) facilitated the gradual dissolution of servile bonds in some areas, while in others it furthered the entrenchment of servitude. We can note that first result in the areas adjacent to the large commercial routes, where the presence of the trader was a given. But in those areas where access to the market was limited to the upper strata of the feudal order, the reinforcement of servitude predominated. Thus, the development of the mercantile economy (with its correlated processes of division of labor and specialization of production) aggravated conditions of servitude—and in the end fostered peasant insurrections. Conversely, the enlargement of the market and increase of distances hastened the differentiating process within urban societies: The producer, no longer dominant in the marketplace, tended toward proletarianization—which in turn led ultimately to urban insurrections. Thus, the social crisis germinated in both rural and urban sectors.

To the extent that they disorganized production, the prolonged and persistent recurrence of these social crises tended in their turn to restrict the very development of commerce. These trends were even further aggravated by monetary depressions, for the European economies relied on the external supply of coinable metals. Such a situation led to a stiffening of competition among various commercial centers and a tendency to close off and dominate principal trade routes. As the principal commercial sector—the commerce in oriental products—was dominated by Italian merchants (mainly from Venice and Genoa), the other trading centers (Flemish, English, French, and Iberian) redoubled their efforts to open new routes.

It was within the general framework of these pressures and as a function of them, that the formation of nation-states took place. The advent of absolutist monarchies (with territorial unification and political centralization) was in fact a response to crisis; or, better, the political redirection was brought on by multiple tensions. The centralized state, on the one hand, effectively promoted the stabilization of internal social order and, on the other, effectively stimulated transoceanic expansion designed to overcome the crises in the various sectors.

In any case, the opening up of new frontiers for mercantilist use and the establishment of new routes across uncharted seas involved unprecedented margins of risk and demanded, moreover, a previous accumulation of capital that would have been far beyond the organizational capabilities of the Middle Ages. The volume of resources to be mobilized, the uncertain profitability, the slow return of investment—all of these made inviable the tackling of the undertaking within medieval forms of mercantile associations. Only a centralized state could function as an organizing center to overcome such a crisis or crises, allowing the gathering of resources on a national and international scale. That was the only reason why a tiny, but precociously centralized Western European state—Portugal—was able to strike out across new routes, opening the way for Europe to circumvent these economic and social crises.

This point also illuminates the contours of European mercantile capitalism as it existed in Portugal during its early modern phase, and makes explicit the previous correspondence between the formation of nation-states and expansion across the seas. Portugal, Spain, the Netherlands, England, and France all sallied forth into the commercial and colonial contest in a measure proportional to their level of internal organization as unitary and centralized states. The rise of such states was an asynchronous process, the occurrence of which varied over both time and space, and each with its own form as a new player in the field of international relations. In its essentials and on the whole, however, this political process emerged from the tensions of feudalism indicated above. The leveling of all as subjects of the crown—which centralized all power in order to delegate it—allowed the disciplining and smoothing of tensions and social conflicts. This step coincided with the simultaneous prosecution of an active mercantilist economic policy, overcoming all impediments to the development of the market economy. Regaining the path toward economic expansion served, in its turn, to alleviate existing social tensions.

In truth, the modern state put in motion—with greater and lesser intensity, enjoying triumphs and suffering frustrations throughout its entire existence—a mercantilist economic policy. It simultaneously prescribed the abolition of internal custom duties and consequent integration of the national market, adopted rigidly protectionist foreign tariffs to promote a favorable balance of trade and the consequent influx of bullion, and established colonies to complement and autonomize the metropolitan economy. The consonance between this political economy and the phase of commercial capitalism subjacent to it was therefore perfect. Likewise, its application strengthened the absolutist state through royal fiscal policies that completed the network of interrelationships. That consonance, highlighted by W. Stark, substantially reduces the validity of the criticisms of mercantilism leveled by later theoreticians (beginning with the classical economists), who built on a conceptual system that largely ignored the historical place of mercantilist doctrine.[6] Absolutism, a society of orders, commercial capitalism, mercantilist policy, and overseas and colonial expansion were all parts of one whole—interacting within that complex that one could call, in keeping with the traditional term, the Old Regime.[7] These were, as a rule, correlated and interdependent processes, all of them products of the social tensions generated through the disintegration of feudalism and the gradual move toward a capitalist mode of production.

During this intermediate phase, in which expansion of mercantile relations tended to eclipse the manorial economy and foster the transition from a servile to a wage labor system, commercial capital commanded all economic transformations. But the mercantile bourgeoisie encountered all manner of obstacles to their maintenance of a rhythm of expanded activities and upward mobility. Thence arose the need for external supports—the colonial economies—to foster concentration in the economic sphere, and, in the sphere of politics, to aid the centralization of power that would unify the national market and mobilize developmental resources. In that sense, the political aspects of the Old Regime—that odd, apparent projection of state power above society—represented the mercantilist bourgeoisie's formula for securing conditions such as would guarantee their own upward mobility and create an institutional framework for the development of commercial capitalism. It meant, at root, simultaneously subordinating all to the king and redirecting royal policy to favor bourgeois progress so that, later, beginning with the French Revolution and from the nineteenth century on out, the bourgeoisie might become—as Charles Morazé would say—"all conquering," modeling society in its own image, according to its own interests, and in keeping with its own values. The strategy was not always explicit on the level of individual consciousness, and it was always fraught with countless difficulties. The history of the process is exceedingly tortuous, allowing Fernand Braudel to speak of the "betrayals" of the bourgeoisie.[8] But amid the contradictions surrounding the development of capitalist expansion and the rise of the bourgeoisie, that underlying mechanism—subjacent to the entire process—is too easily overlooked.

Within and inseparable from this context, one is able to focus on European transoceanic expansion and the formation of the New World colonies. Modern European colonization reveals itself in the first place as an intensification of a purely commercial expansion. The American lands were discovered during the course of opening up new trade routes for European mercantile capitalism, and the first activity undertaken was the barter for raw materials with the aborigines. Actual settlement was initially spurred by the need to guarantee possession in the face of disputes over the dividing up of the new continent. Production for European markets was a way of making these new domains viable. Almost imperceptibly, the shift from commerce to colonization was accomplished, although this development involved a new form of activity, a fact that did not escape the notice of the keener observers of the day.[9] In effect, during the transition from commerce to colonization there was a shift from the emphasis on commercializing goods produced by established societies toward the production of merchandise by a new society. So it was that the occupation, settling, and development of these new areas, as well as their integration into the European economy, were undertaken. In this way, economic activity eventually went beyond the field of circulation of merchandise to promote the implantation of complementary, extra-European economies; that is, it entered the orbit of production proper.

Notwithstanding such fundamental differences or the new dimensions that the colonizing activity acquired as it transcended its initial engagement in overseas trade, colonization kept within its essence that sense of commercial endeavor from which it emerged; only the nonexistence of marketable goods led to their production, and that, in turn, prompted colonizing activity. That was how the new areas adjusted themselves within the framework of European economic needs. Modern colonization, therefore, as incisively indicated by Caio Prado Júnior, has an essentially commercial nature: to produce for the foreign market and to furnish the European economies with tropical products and coinable metals—that, at root, was the "meaning of colonization."[10] If we combine this formulation—the commercial nature of early modern colonial endeavors—with the observations made earlier about the Old Regime—the intermediate stage between the disintegration of feudalism and the development of industrial capitalism—then our search for the real "meaning" of colonization will be well under way.

To summarize so far: The formation of a market economy in Europe began through the occasional sale of surpluses from premercantile production. To the extent that this commercialization became permanent, that sector of society dedicated to the acquisition of capital through the circulation of economic goods gained prominence. Bit by bit, and as a function of the same process, production exclusively for trade was introduced, as was specialization in production. Hence, the accumulation of commercial capital, the division of labor, mercantilization of economic goods, and specialization of production were all correlated processes involving a general raising of the economic level. The concentration of commercial capital and formation of the mercantile bourgeoisie were thus two facets of the same process.

Theoretically, the transformation was self-stimulating without limits. Historically, however, the launching of this process was connected to a concrete reality—the feudal system. From that reality derived the growth of social tensions beginning with the formation and expansion of a mercantile sector within the framework of the feudal economy; thence, also, the continuous political readjustments that channeled those tensions. The latter part of the Middle Ages represented a critical time because of these tensions and adjustments. The processes unleashed during the overcoming of this crisis included the creation of a unitary central state as executor of mercantilist policy, transoceanic and colonial expansion, and the creation of instruments to accelerate the development of a market economy, providing incentives for capital accumulation.

Eventually, capital was introduced into production. The expansion of the market economy, with its correlated processes of division of labor, specialization of production, and consequent raising of the level of productivity, only acquired a self-developing dynamic, however, with the emergence of mechanized industrial production. From artisanship to manufactures—where capital and labor were already dissociated— and on into the factory system there unfolds the process of forming capitalism, covering the entire period from the end of the Middle Ages through its completion with the Industrial Revolution.

As long as the last step remained to be taken, however, the commercial capitalist economy—and hence the rising mercantile bourgeoisie—still lacked sufficient internal growth capacity. The capitalism that resulted from the pure and simple workings of the market did not produce the transition to the decisive component—the mechanization of production. Hence arose the need for supports external to the system, inducing an accumulation that, because it was generated from outside the system, Marx referred to as original or primitive.[11] Hence also, toward the end of the eighteenth century, came the political and social tensions caused by the confluence of an entire complex of stimuli. Mercantilism was, in essence, the confirmation of the system's need, and the mercantilist colonial structure was one of the economy's fundamental parts; a lever for the emergence of modern capitalism. Contrary to what Max Weber believed, colonial exploration was among the decisive elements in the creation of the prerequisites for industrial capitalism.[12] In fact, the taking of that ultimate and decisive step in the transition to the capitalist order required, on the one hand, ample accumulation of capital on the part of the business class, and, on the other, a steady expansion of the consumer market for manufactured goods. To be sure, both of these prerequisites were generated by the process of developing a market economy proper. As we have seen, however, the pure form of this mechanism ran into immovable obstacles, the overcoming of which summoned forth the colonial system and mercantilist policy.

When examined in this context then, the colonization of the New World in early modern times reveals itself as a cog in a system, as the instrument of primitive accumulation within the era of merchant capitalism. That which, at the outset of these considerations, appeared to be merely a project, an idea, now presents itself as intimately consonant with the actual historical process of forming capitalism and bourgeois society. One also appreciates the deeper dimensions of colonization: both *commercial* and *capitalistic*—in the sense of being an essential formative element in modern industrial capitalism.

We can now finally comprehend the colonial system in its multiplicity of connections. This set of mechanisms—whether examined in the rules of political economy or in actual economic relations—integrates and links colonization with the central European economies, representing an immanent and subjacent reality underlying the colonizing process and constantly steering its outcome. It was not merely a common denominator present in all concrete manifestations of the historic process, but, rather, the structural determinant itself, that component that makes it possible to comprehend the sum total of manifestations, making reality intelligible—the element that finally explains and defines the others—but is not itself defined by them.

In the real world, colonization does not unfold strictly within the frame of the colonial system. Such systems never present themselves, historically speaking, in their pure form. Albeit coeval, the colonization of New England took place outside of the mercantilist colonial system's defining mechanisms. As we have indicated elsewhere, specific factors—the English politico-religious crises, occurring as they did during the formation of the modern English state—gave rise to a particular form of transoceanic expansion: settlement colonies (to use the terminology consecrated by Leroy-Beaulieu).[13] Based predominantly on small properties, their production mainly responded to the colonies' own internal consumption. In contraposition to these, the category of colonies Leroy-Beaulieu called exploitation colonies, whose production organized itself in terms of large slaveholding properties (as, for example, in Brazil), had economies completely turned back toward the external, metropolitan market. During the course of our exposition, these colonies have assumed new and easily discernible dimensions: the exploitation colonies were those best adjusted to the framework of the colonial system. The settlement colonies were those that remained relatively peripheral to that system. Yet in order for the explanatory scheme we're developing to be true, what with both types of colonies being generated from within the same colonizing process, the whole, and hence the settlement colonies as well, must be understandable in terms of that system, rather than the other way around. We will get to that in due time.

Similarly, if colonial Brazil was framed along the lines of the Old Colonial System as an exploitation colony, that does not mean that all colonial manifestations of Portuguese America directly expressed that mechanism. Rather, the mercantilist colonial system's mechanisms constitute elements fundamental to the whole, in terms of which it ought to be analyzed. At this early point in our analysis, however, we are trying to render explicit the basic category (the colonial system) in order then to comprehend its mechanisms and to study its crisis on a structural level. Naturally, we will further on have to take up some of the elements advanced here in order to redefine the positions of metropolitan Portugal and colonial Brazil within the entirety of the system. Thus, our analysis will become more concrete as we go along.

The Metropolitan "Exclusive"

Let us now examine the functional mechanisms of mercantilism's Old Colonial System. The essential element underlying this system lay within the regime of commerce occurring between metropolises and colonies. In reserving to themselves exclusive overseas commercial rights, the European metropolises organized an institutional framework of relations that stimulated the primitive accumulation of capital in the metropolitan economy at the expense of the peripheral colonial economies. The so-called colonial monopoly, or, more correctly and using a term of that era, the metropolitan "exclusive," made up the systemic mechanism par excellence through which colonial expansion was adjusted to the economic processes of European societies during their transition to capitalism.

Commerce was in fact the nervous system of the Old Regime's colonization. That is, the occupation, settling, and development of new areas were undertaken in order to foster mercantile activities. Along with the increasing mercantilization of various productive sectors in Europe and a widening circulation of merchandise there, colonial production on the peripheries of the European dynamic center was also mercantile production closely connected to the major international commercial routes. That alone would suffice to indicate the significance of colonization as a stimulating adjunct of mercantile capitalism. Colonial commerce was the metropolis's exclusive commerce—and it generated huge profits, an aspect that completes the characterization. Despite all the variations that colonial-metropolitan relations underwent during the course of the sixteenth through the eighteenth centuries, commerce was always the basic matrix of these relations, encompassing situations that diverge from this typical process only as variations arising from special factors or circumstances. The commerce that developed through the opening up of new maritime routes at the beginning of the early modern period was effectively exclusive. Throughout the course of Portuguese expansion during the 1400s, exploitation of commerce along the African Atlantic coast was the prerogative of the king alone, that is, of the absolute monarchist state, although he could delegate it to others.[14] Among these recipients figured the Order of Christ, personified by its Grand Master the Infante Prince Henry, as well as private and even foreign merchants. Still, the regime's basic exclusive principle never strayed, nor did the underlying mechanisms ever stop functioning.

The rounding of the Cape of Good Hope presented the Portuguese with the possibility of opening up commerce along the African and Asiatic shores of the Indian Ocean. These worthies then erected a military-political complex, the Portuguese vice-royalty of India, in order to exclude the Mussulmans and, through them, the Italians from participation in those mercantile activities. In other words, a coercive apparatus was organized in order to guarantee the monopolization, and hence the high profitability, of the route around the Cape. It amounted to an attempt to block off the entrances to the Red Sea and the Persian Gulf. Commerce was thus organized as a royal monopoly with the King of Portugal as the sole agent. Resources for the commercialization of oriental products were mobilized through the agency of the state. However, the scarcity of capital in Portugal alluded to earlier led the Portuguese crown to resort to foreign capital, especially that of Flanders, and so to transfer the European commercialization of oriental products into Antwerp's hands. This lent these Flemish entrepreneurial groups greater control over European prices. Their manipulation of prices, in turn, nudged them ever more into the capacity of financiers and creditors for royal undertakings; leaving the Portuguese crown to assume the not-insignificant risks of transportation. In addition, the profits kept by the crown, on being integrated with the other sources of royal treasure, were not necessarily reinvested in oriental dealings, but were often diverted to attend to other expenses incurred by the Portuguese state. Thus, the "monarchic capitalist" scheme ended up frustrating the rationality of commercial enterprise in oriental products, greatly weakening the Portuguese position on the whole, and finally giving rise to bankruptcies and failures.[15]

One may nevertheless observe that such distortions took place on the level of distribution of profits generated through monopolized commerce. The essential aim was that there not be competition among buyers in the Orient, which would reduce the profits to their normal value in commercial transactions. The Portuguese royal monopoly thus guaranteed conditions favorable to the European economy in general, thereby promoting the acceleration of mercantile capital accumulation. Within the actual workings of the system, however, the bulk of its advantages were transferred outside the kingdom. With that, Portuguese domination in the Indies weakened in the end, resulting in a diminution of the volume of its commercial activity.

The Portuguese retreat facilitated the entry of the Dutch in the early seventeenth century. Despite the Dutch War of Independence (beginning in 1579 with the Union of Utrecht) and the annexation of Portugal by Spain in 1580, strong Dutch participation in oriental trade continued by way of Lisbon. In 1585, however, the year the Spaniards captured Antwerp, Dutch ships were apprehended in Lisbon. Even then, Dutch entrepôts were so bound up with and important to oriental commerce that Phillip II made a last-ditch effort to avert the breaking off of trade relations. Still, by 1598 all trade with Holland was prohibited with the passage of the impoundment decree—described by Grotius as a "barbaric edict."[16] By this time the Dutch were already organizing the launching of direct commercial relations with the Orient. They mobilized their resources, and, in April of 1595, they made their first voyage. Although the compensation from this effort was rather paltry, the route to India, at least for the Dutch, was now open.

The Netherlands, within the context of the European economies, now found themselves in a peculiar position. Ever since the Middle Ages the region had figured prominently as one of the most active centers of development for the European merchant economy. Bruges, in the Lower Middle Ages, and Antwerp, beginning with the sixteenth century, had been economic and financial centers rivaling the Italian cities.[17] Flemish wealth arose as a consequence of its position as an entrepôt for transfer and distribution of products from various European economic regions; in sum, of its carrying trade. Its economic policy tended toward the lightest possible regulation in a spirit of grand liberalism in order to attract, and then redistribute, merchandise from all areas. In this manner and rooted in this tradition, in the latter part of the sixteenth century the Dutch organized many autonomous companies in efforts to trade directly with the Orient. Between 1595 and 1602 some ten companies were formed, deploying sixty-five ships. Few enjoyed success. For the majority, the results were disastrous. They ended up competing with one another in the purchase of oriental products. Their situation was further aggravated by the peculiar conditions attending such longdistance commerce in a region made hazardous by the monsoons of the Indian Ocean.

The need for a change in economic policy with respect to oriental trade became obvious. The Amsterdam Company, having done fairly well, petitioned the Estates General for a monopoly concession over this sector of Dutch trade. The petition was denied, and this action touched off heated debates and polemics until finally the monopoly was imposed with the charter of the East Indies Company (March 20, 1602), guaranteeing it exclusive mercantile activities in the Orient (between the Cape of Good Hope and the Straits of Magellan), with the right to sign agreements, name functionaries, and the like.

This Dutch experience is thus highly significant for the explanation of the mechanism we are analyzing, effectively offering a counterproof. The free practice of overseas trade, having been tried, was found ineffective for the requirements of European mercantile capitalism in need of external stimuli; the failure of the attempt led, in practice, to the adoption of a monopolist scheme.

It was within the context of monopolist overseas exploitation that colonial production was begun, and from it the commercialization of products generated within the New World. The earliest exercise in colonization proper took place, as is known, in the Atlantic islands—particularly on the island of Madeira. The cultivation of cane and production of sugar were introduced to these islands during a phase when the resources of the small entrepreneurial kingdom of Portugal were concentrated on circumnavigating Africa. From the outset, therefore, this colonizing effort depended on the participation of foreign resources and capital. The Genoese, it seems, were foremost among those connected with setting up this new economy, an economy by means of which the Venetian monopoly in certain products was being broken. Then, with the breakup of that monopoly, there came an attendant increase in consumption of sugar that was eventually serviced by the Dutch; by the last quarter of the fifteenth century a state of overproduction had clearly been reached, bringing on the restrictive measures instated by Manuel I, who in 1498 fixed maximum production at 120,000 *arrôbas* annually, of which 40,000 went to the Netherlands.[18]

Even earlier, however, in 1482, amid the airing of numerous complaints in the Portuguese *cortes* of Évora about the economic activities of "foreign residents, such as the English, Flemish, Castillians, and

Genoese" who caused "great damage to the people of thy kingdom" and "bring great impairment of thy taxes," the situation on the islands was being seriously criticized. Remembering that Prince Henry, the "inventor" of those islands, had not condoned the presence of foreigners there and had insisted that island products come first to the kingdom, paying the duties and yielding freight rates for national ships, only then being shipped abroad, the *cortes* decried the concessions that now permitted the residence of foreigners on the islands, concessions that resulted in the direct transport of goods outside of the kingdom (during 1480, twenty Spanish ships and forty or fifty from other nations had done so). The alleged loss of royal duties "both on entering and leaving" and large losses borne by the people prompted a petition to "determine, your Lordship, and uphold that foreigners not be countenanced as residents on said islands, nor allowed to load ships outside of the kingdom, and that all sugars and other merchandise come to Lisbon or to other ports in thy realm where they may be unloaded and from thence carried by whom you approve to whence they will, rendering payment upon departure." This "will be a great advantage to thy revenues and a great benefit for the common good," for otherwise there will be a "loss of shipping" in Lisbon and "other places in Portugal."[19] The interests of Portugal's mercantile bourgeoisie could not have been more clearly put than in these complaints.

The proposal amounted to an appeal for the restructuring of the Atlantic islands' colonization into the framework of monopolist overseas exploitation. Once again, the exclusivist logic of colonization is made clear. As a consequence of these demands, the residence of foreigners on the colonized islands was indeed forbidden, allowing those residing there a period of one year within which to leave. One observes, then, the policy astutely followed by the Portuguese crown: free trade in the initial stage to stimulate the entry of resources and capital for the installation of colonial production; restructuring into the exclusivist system once the peripheral economy had begun functioning fully.

The implantation of the sugar economy in Brazil featured, after a fashion, a repetition of the process. During the first purely predatory economic contact, nothing beyond the commercialization of natural products—the bartering for brasilwood with the aborigines—was tried. Such commerce was at the outset considered a "monopoly" of the crown, which delegated it to the recently converted New Christian entrepreneur Fernando de Noronha or Loronha. It amounted, then, to a simple unfolding in the Americas of the regime already applied to African and Indian commerce. During the transition to colonization, that is, in the introduction of cane cultivation and sugar refining, the king resorted to private entrepreneurs through the concessions of captaincies whose charters still perpetuated the principle of royal monopoly. We know that few of these licensees enjoyed success—as did Duarte Coelho in Pernambuco—in the difficult business of installing costly agro-industries in Portuguese America. None of this gainsays the hypothesis advanced by Celso Furtado that during this arduous phase recourse had to be made to foreign capitalists, especially Dutch, as they were already heavily involved in European sugar dealings—although monographic studies have not yet fully validated that affirmation.[20] It is nevertheless correct that during this early phase the commerce in goods was relatively free. There are even reports of licenses for direct commerce to foreign ports.

The sugar economy expanded in this manner so that in the decade of the 1560s Brazil already boasted 60 mills, producing some 180,000 *arrôbas*. The greatest upswing, however, occurred during the final quarter of the sixteenth century and the first decade of the 1600s. Calculations indicate that in 1610 there were already some 650 mills in existence, the surge in production corresponding with an accentuated upward movement of prices. The sugar-price curve for Lisbon reveals remarkable increases, but, as noted by Frédéric Mauro, in Brazil prices remained nearly stable.[21] This was because at the opening of the prosperous phase, King Sebastian decreed (February 3, 1571) the exclusivity of Portuguese shipping in the then-flowering colony. One notes the coincidence between the decree and the boom of the Brazilian sugar economy; it was, in essence, the framing of the new peripheral economy within the structural lines of the colonial system. One also notes that the end of the sixteenth century was marked by intensified repression of foreign commerce and multiplying seizures of foreign ships.

It is quite true that external pressure was mounting at the time and that the Spanish monarchy [to which Portugal and thus Brazil were now subjugated] was struggling with enormous financial difficulties. This led the King of Spain and Portugal, despite new prohibitions (for example, of February 9, 1591), to concede some special licenses—to the point of permitting direct regular trade between Brazil and Hamburg, which sent nineteen ships between 1590 and 1602. But after that time, all indications are that direct voyages to non-Portuguese ports ceased. Let us immediately emphasize here, however, that these licenses in no way altered the fundamental mechanism we are exploring. Such concessions did not imply the establishment of competition among buyers. One may legitimately affirm, as substantiated by price records, that along with the rapid growth of the sugar economy we also witness its reshaping along the lines of the colonial system. Prices rose little in the colony, but increases were sharply noticeable in the metropolis, that is, excessive profits were generated—monopolistic profits—that accumulated among metropolitan merchants.

Repercussions of the intensifying clashes in the war between Spain and Holland were clearly felt by the commerce to Brazil. Repeated prohibitions attest to the increasing pressure of smugglers. Thus, in January of 1605 restrictions were imposed again on foreigners seeking to enter Brazil or, rather, on their ships, obliging the petitioners to submit their proposals to the Council of the Indies itself, whose president would sign a passport if the petition were conceded. New restrictions were enacted on March 18, 1605: no foreign ship of whatever nationality could go to Brazil, India, Guinea, and the Atlantic islands, nor to any other lands either discovered or yet to be, excepting only Madeira and the Azores; all foreigners residing in overseas Portuguese territories were required to move to Portugal within a year or face the penalties of death and confiscation of property. That this legislation in itself was naturally impotent to maintain the Portuguese monopoly—which really depended on military conditions to oppose pressure from Holland—does not make the adoption of an exclusivist commercial regime any less patent. These principles were incorporated into the *Philippine Ordinances* of 1603.[22] Smuggling certainly did not cease, but the decision on the part of the low countries to set up a special company to deal with the West Indies and to organize the military occupation of the sugar-producing northeast of Brazil shows that smuggling did not suffice to satisfy the expansive needs of the Dutch economy.

The restoration of Portuguese independence from Spain (1640) marked a temporary diminution of Portuguese overseas exclusivism. The internal political situation faced by the new government and the weak Portuguese position internationally explain the concessions made to Holland and England in exchange for alliances in the struggle against Spain. It was exactly because the Portuguese colonization of Brazil was by that time already structured within the functional lines of the colonial system that the concession to foreigners of the right to participate became a powerful coin with which metropolitan Portugal could bargain in its anti-Spanish schemes. The concessions contained in the treaties with England (1654) and Holland (1641) amounted at root to the participation by these countries in the fruits of exploitation by the Portuguese colonial system.

From another angle and along similar lines, the Portuguese government sought to organize its system of overseas control more efficiently, mainly through the creation of the Overseas Council, which came to oversee all colonial activity. It was an effort to exercise as much control as possible over concessions already granted. Similar considerations prompted the institution of the monopolistic General Commerce Company for Brazil in 1649.

From then on, amid increasing colonial competition among the powers, the Portuguese crown struggled diligently to minimize the breaches of its colonial monopoly. In a petition for redress of grievances in 1672, Portuguese merchants loudly sought reparations, for they were finding the Brazilian markets saturated when their ships arrived there. The Edict of November 27, 1684, forbade ships departing Brazilian shores from proceeding to any ports other than Portuguese ones. The Royal Order of February 8, 1711, in the same vein, established that foreign ships (those admitted by treaty) could only go to Brazil in the official fleets or in cases of forced landings, and prescribed rigorous penalties for infractions. One measure followed another, gradually canceling out the concessions to foreigners, until their legal presence was reduced to cases of forced landings only. The process culminated with the Edicts of June 19, 1772, and December 12, 1772, which, revoking the concessions of 1765 and 1766, prohibited all intercolonial commerce while stating that it was "a generally accepted maxim constantly practiced among all nations, that Commerce and Navigation to the Colonies should be made from the Capital, or Dominant Metropolis, and not among the colonies themselves."

If we now examine, albeit more succinctly, the economic relations established during the process of Spanish colonization in America, we find ourselves faced with the same principles and the same mechanisms. The Castilian colonial enterprise surfaced initially as an exclusive business of the crown, associated with Christopher Columbus. The widening of the enterprise, while necessarily reducing the audacious discoverer to a position of insignificance, consolidated the royal monopoly, which naturally embraced its Castilian subjects. In reality, beginning with the institution in 1503 of the Sevillian Casa de Contratación, all legal commerce with Hispanic America came to be done through the Andalucian port. This was the single-port regime that was only altered at the end of the eighteenth century under the enlightened despotism of Charles III's ministers. That important Sevillian organ, despite having been subordinated since 1524 to the Royal and Supreme Council of the Indies, superintended all colonial traffic, vigilantly safeguarding the monopoly. External pressure, in the form of piracy and harrying unleashed by rivals who by the first half of the sixteenth century were already equipping themselves for overseas competition, brought on a stiffening of the regime. Already in 1543 and more so between 1564 and 1566, navigation was limited to convoys—"fleets of galleons"—sailing during specific seasons along predetermined routes to a few privileged ports in the Americas from which proceeded the distribution of metropolitan products. Veracruz in New Spain, Cartagena on Tierra Firme, Panama and Porto Bello on the Isthmus were the privileged centers. One result, for example, was that the provisioning of Buenos Aires and the Río de la Plata River basin had to be done exclusively by way of the Pacific. The result of the Sevillian merchants' monopoly or that of their associates was, in Professor Céspedes del Castillo's synthetic formulation, "a regime of large profits [for Spain] that determined a regime of high prices in the Indies."[23] There is no doubt that Spanish colonization, like the Portuguese and the Dutch, was organized along the lines of the mercantilist colonial system, which tended to set up mechanisms designed to accelerate primitive capital accumulation in Europe. The fact that Spain proved unable to reap its advantages—that they ended up being transferred to rival powers—stems from conditions peculiar to the metropolitan situation.

Clearly, a regime given to such single-minded inflexibility had to provoke immediately the defiance of rival powers, who hastened to provide incentives for smuggling into Hispanic America. Beginning in colonial Brazil itself, an enormous trade in illegal commerce soon developed into the Río de la Plata region, especially during the period of the Iberian Union. Englishmen, Frenchmen, and Dutchmen gave the exclusivist Castilians no respite until, in the seventeenth century, having ensconced themselves in the Antilles, they erected competing colonial economies while setting up entrepôts to stimulate traffic in contraband to the Indies of Castile. The Spanish system offered its enemies several enormous flanks, the most important of which was certainly the traffic in black slaves to the Hispano-American colonies. The difficulties in setting up African posts led the Spanish crown to subcontract the provisioning of their colonies to foreign merchants through the *asiento*. Competition for this highly lucrative overseas traffic was especially violent; the Portuguese, Dutch, and French successively controlled the *asiento*, which was finally awarded to the English through the Treaty of Utrecht (1713). It is well to keep in mind that contraband does not negate the reality of the colonial system. What the entrepreneurs from rival powers sought was precisely the reaping of advantages inherent in this system. This truth is underscored by the fact that the colonial policy of these same powers (Holland, France, England) never wavered in its adherence to that same policy that had crystallized during the first phase of transoceanic expansion.

Overseas competition, undertaken at the outset on a purely commercial level, soon unfolded, as we have seen, into a veritable colonial competition with the installation of the English, French, and Dutch colonies. We have already dealt, albeit summarily, with the Dutch experience: their commitment to the establishment of direct trade routes to the Orient, an effort that led to their organization of a monopolistic trading company. Dutch dominance in the Orient soon transcended the stage of a purely mercantile action. Occupation of large islands such as Java and Sumatra gave way to a colonizing effort, shifting to the production of spices. The undertakings were still executed, however, within the monopolistic framework of the powerful East Indies Company. Their scheme for western expansion—to the West Indies—was no different: being set in motion through the agency of the West Indies Company, a clone of the first. Under its guidance and control, in addition to the temporary domination of the Brazilian Northeast, it also promoted the occupation and exploration of Surinam and Curaçao.

English maritime expansion, for its part, also adhered closely to formulated mercantilist principles. We've already mentioned Thomas Mun, defender of the British East Indies Company. He was followed by an entire dynasty of theoreticians (Josiah Child, Joshua Gee, Malachy Postlethwayt, to name a few of the most representative) who carried mercantilist doctrine to the highest degree of refinement, and, within it, the theory of the colonial system. English colonization itself exhibited the most variegated hues and aspects, at times assuming inconsistent forms, but Great Britain nevertheless prevailed in colonial competition to become, in the nineteenth century, the imperial power par excellence. In the early phase (during the sixteenth century), it embarked, along with Holland and France, on a parasitic venture: the plunder of Spanish colonial trade. The beginning of the seventeenth century marked its actual colonial expansion in several directions, including channeling dissident groups formed during the political and religious crises—in the midst of which the formation of the modern English state took place—into North America. This action gave rise to a pattern of colonization—the settlement colonies—different from that of the general European framework. But, finally, in the middle of the seventeenth century, "plantations" were set up in the Antilles.

The English "Old Colonial System" was given legal expression by the famous Acts of Navigation. The Act of 1651, under Cromwell, established that American, Asian, and African products could be carried to England only by English ships or those of English colonies. European products were admitted only aboard English ships or those from the country in which the products originated—a clause that excluded Dutch intermediary carrying trade. Exceptions were made, as for Italian silk, which was admitted from Flemish ports, and for products from the Spanish and Portuguese colonies, which could be imported from Spain. Note that the exception indicates a harmony of interests with Portugal and Spain, for England effectively had an interest in these imports—imports that, in a complementary manner, made it possible for British manufactures to reach Latin American markets by way of the metropolises. Another aspect worth mentioning in the Cromwell Act was the integration of measures encompassing overseas colonization as a whole (products from America, Asia, and Africa) in a single act, along with regulations concerning English trade with other European powers—doubtless, indications of the coherence of mercantilist policy, of which the colonial system forms only a part, albeit a central one.

The Act of 1660, enacted during the Restoration, underscores the persistence of English mercantilist policy in the wake of Cromwell's fall from power. It defined as English those ships whose captain, along with three-fourths of the crew, were English. It also specified that products from English colonies could be carried only by those ships, reaffirming an earlier ruling. Indeed, it established the "enumerated articles" that could leave British colonies only if bound for either England or other English colonies. These were the staple products of transoceanic commerce: sugar, indigo, cotton, tobacco, and timber. Three years later the "Staple Act" (1663) forbade the colonies from importing on ships that had not docked at English ports, opening exceptions for Madeira wine, French salt, and Scottish and Irish horses. Another Act, in 1673, taxed the enumerated articles that circulated from one colony to another. The system was reaffirmed in 1696 in an Act purporting to "prevent fraud and regular abuses in colonial commerce" (i.e., plantation trade).[24]

For France, the early phase of maritime expansion was characterized, as for England and the United Provinces, by piracy and harrying. Some ill-fated attempts at overseas occupation and settlement did occur in the meantime, but only under Richelieu (1624–1642) was expansion given new momentum, yielding its first rewards. Monopolistic new companies were created to further overseas expansion. The results achieved by such companies in New France (1627), the islands of America (1635), Senegambia (1641), and in the Orient (1642), although not brilliant, nevertheless established the first bases. Under Jean Baptiste Colbert (1669–1683), French mercantilism—or Colbertism, as it came to be called—was structured as a wide-ranging plan, simultaneously affecting all sectors of the national economy and then providing the most complete example of the simultaneous application of mercantilist policy. Overseas and colonial expansion was thus organized within the framework of the monopolist scheme. Colbert reapplied Richelieu's policies, reorganizing the privileged companies and gave them new and decisive momentum. Accordingly, the East Indies Company, the West Indies Company, and the Senegal and Guinea Companies retained their exclusive rights over the various areas of French overseas commerce (trade in oriental products, colonial products, slave trade, etc.), and it was within this context that French colonization took root.

By the second half of the seventeenth century, as colonial competition among the European powers was defined and crystallized, we find overseas development organized within a commercial regime that, despite minor variations and fluctuations, everywhere displayed at root the same fundamental mechanism. All of the processes operated within the same basic exclusivist system. The tensions of competition, the struggle among the powers, the smuggling—none of this negated that system. If we visualize it in conjunction—on one side, European mercantile capitalism during its rapid expansion phase and, on the other, the peripheral colonial economies—we restate, in essence, the system of exploitation of the latter by the former. The conflicts among the various Old World nations arose precisely over the reaping of their advantages and the redistribution of commercial and colonial profits.

We focus, then, on the basic mechanism of the commercial regime; the axis of the mercantilist colonial system. The metropolitan "exclusive" over colonial commerce consists, in sum, of the reservation of the colonial market for the metropolis, that is, for the metropolitan commercial bourgeoisie. This was the fundamental mechanism, generating excess profits, colonial profits. Through it, the central metropolitan economy incorporated the surplus products of the ancillary colonial economies. By retaining exclusive purchase rights for colonial products, merchants from the homeland were able to depress colonial prices to a level below which the continuation of the productive process would have become impossible, that is, toward the level of production costs. On the other hand, resale in the metropolis, where they enjoyed exclusive selling rights, guaranteed them excess profits too; thus, they benefited on both sides—in purchase and in resale. This promoted, on the one hand, a transfer of real income from the colony to the metropolis, as well as the concentration of this capital among the merchant classes connected to overseas commerce. Conversely, the possessors of exclusive sales privileges over European products shipped to colonial markets were able to resell in the colonies at high prices, prices above which consumption would have become impracticable. Here, then, was a reiteration of the same mechanism stimulating the primitive accumulation of capital by merchandisers in the homeland. Further on, we will analyze other aspects that will enable us fully to comprehend this process of original accumulation in all its dimensions; in the meantime, let us simply state that we will argue that the socioeconomic structure being organized in the colonies—slave-based production and the ensuing concentration of income in the dominant strata—was what made possible the functioning of the entire system.

But here let us further particularize the mechanism whose exclusivist essence is described above. The metropolitan "exclusive," just as the subordination of the colony, could assume many gradations, complicating the scheme in many ways. It is true that the "exclusive" over maritime trade could, in the extreme case, belong to a single businessman; as in the case, for example, of royal monopolies granted to special contractors or even retained by the Portuguese king himself, as was true during the earliest phase of oriental commerce. In this circumstance, the sole entrepreneur retained an exclusive right over the purchase of external products, that is, of their procurement in the colonial market (here we speak, in technical terms, of a "monopsony"). He also retained, naturally, an exclusive right over the sale of the product in the central economy (a "monopoly," technically speaking). The most common arrangement, however, was for exclusive access to colonial commerce to accrue to the metropolitan mercantile business class as a group. In the colony, this group retained exclusiveness in procuring colonial products (that is, "oligopsony"), as it did in the sale of European products in the colonial markets ("oligopoly"). The situation typifying the colonial system, were we to classify it technically, would then be that of an "oligopsony-oligopoly" or "bilateral oligopoly." Intermediate between the sole agent and the entire class of metropolitan merchants, the exclusiveness could remain restricted to a more specific group of metropolitan businessmen, as in the Spanish single-port system, which favored those merchants linked to Sevillian commerce. Colonial trading companies also figured in this intermediate position: for in reality they included only a fraction of the metropolitan merchants.

In the metropolitan markets the situation could, in its turn, also vary. If the group connected with overseas commerce sold colonial products under conditions of monopoly or oligopoly (naturally, at high prices), this promoted a transfer of income from the population of the homeland as a whole to the entrepreneurs involved in colonial commerce. If they resold these products in another country under the same conditions, the transfer of resources was made from outside the national borders to the inside, always concentrating in the same privileged caste; if, however, the same was done under conditions of competition with other nations, this avenue of accumulation declined or could go to other nations. Likewise, the purchase of European products for provisioning the colony could be accomplished under more or less favorable conditions. If the products for provisioning the colony were acquired outside the metropolis, as when the metropolis failed to provision the colonies itself, this avenue of accumulation naturally tended to become blocked.

Some objections to this line of interpretation can still be made. They involve those mechanisms that operated throughout the span of the entire Modern Era, and that, according to some authors,[25] would counteract the functioning of the system: treaties conceding overseas commercial advantages to other powers, licensing of foreign merchants, and, finally, contraband. In our view, nevertheless, such occurrences do not invalidate, but rather confirm our analysis. In fact, such licenses and concessions presuppose that the mechanism of colonial exploitation generates excess profits. Otherwise, what would one really be conceding? If a monarch in need of financial resources eventually sold licenses to foreign merchants, or if a metropolitan state, because of political necessity, permitted merchants from other nations to trade in their colonies (as did Portugal, in the wake of the Restoration)—what was really happening was a transfer of the advantages of the colonial system's economic stimuli. This did not establish real competition. It was, rather, the very possibility of more highly lucrative commerce that made such licenses and concessions so desirable, to the point of wars being fought to obtain them.

Contraband, to be sure, involved a more complex situation, but, as concerns our analysis, is confirmatory nonetheless. It is obvious that smuggling always involved serious risks: imprisonment, confiscation of merchandise and ships, and so on. What could move merchants to take such risks and apply themselves to illegal trade—if not the prospect of colonial superprofits? Contraband, then, also presupposes the basic mechanism rather than negates it. In order to find an area for his activities, the smuggler must certainly offer somewhat better prices for colonial products, as well as offer European products at lower prices than metropolitan merchants; but never at a level implying perfect commercial competition, for otherwise, what would compensate him for the high risks? Capital would then be channeled to other areas of equal profitability and smaller risk. So it seems certain that although smuggling may have been a palliative, it did not imply the end of the system. The basic exclusivist mechanism always persisted as the explanatory element of all this movement.

In sum, licenses, concessions, contraband, all seem to us phenomena best relegated to the area of disputes among the various European metropolises aimed at appropriating the advantages of colonial exploitation. And this exploitation operated throughout the system, that is, in all the relations between the European central economy and the peripheral colonial economies. None undermined the essence of the system of colonial exploitation. They were all variations of its fundamental element.

In the end, the regime of colonial commerce—that is, metropolitan exclusiveness over colonial commerce—constituted throughout the sixteenth, seventeenth, and eighteenth centuries the mechanism through which the metropolitan merchants appropriated the excess profits generated in the colonial economies. Accordingly, then, the functioning of the colonial system impelled the primitive accumulation of capital within the framework of European mercantile capitalism. With such a mechanism, the colonial system adjusted the process of colonization to its own purpose in the economic and social history of the modern era.

The Colonial Economy, Slavery, and the Slave Trade

The analysis we have been sketching out of the Old Colonial System cannot be complete without studying, however summarily, the type of economy organized in the colonies. We've already seen that the guidance provided by the broad outlines of the colonial structure is indispensable to an understanding of all the broader mechanisms encompassed in overseas exploitation. By the same token, the starting point for the characterization of the colonial economy is the larger meaning of colonization and the basic mechanisms underlying metropolitan-colonial relations. European expansion and the organization of the New World's productive activities effectively proceeded from, and as a function of, that underlying purpose. To repeat, the occupation, settlement, and economic development of the new areas developed within the framework of Old Regime commercial capitalism and as a function of the mechanisms and adjustments inherent in that formative phase of modern capitalism. Essentially, European colonial expansion developed according to a fundamental impulse generated by tensions emanating from the slow transition to industrial capitalism in Europe. The acceleration of primitive capital accumulation is, then, the direction of the motion—not present in all its manifestations, but immanent in the entire process.

In this sense, colonial production was necessarily oriented toward those products capable of fulfilling the Old Colonial System's function within the context of merchant capitalism: commodities saleable in the central economy for which potential or manifest demand could be found in European society. These were mainly tropical products: sugar, tobacco, cotton, cocoa, indigo; or raw materials such as furs for luxury clothes, dye-bearing woods, and so on. Beyond those, naturally, they produced coinable metals, lest the expansion of the market economy be constrained by scarcity of currency.

It bears repeating also that the earliest attempt at colonization, in the Atlantic islands, began very early under the direct leadership of Prince Henry, who sent over the first settlers. The initial idea seems to have been to populate these strategically located islands in order to maintain possession of them, while at the same time trying to keep Portuguese routes and discoveries secret. This resulted initially in the organization of an economy oriented primarily toward the pioneers' own consumption, albeit with a little exportation to the metropolis of much-needed cereal grains. Not long thereafter, however, the island economy turned toward the external market, aiming toward Portugal and, soon after that, toward the European market in general. The introduction of the sugar industry, especially on Madeira, and its rapid diffusion, led, little by little, to productive activities tying into an expanding European commerce. With the development of a sugar economy in Brazil, viticulture, which had begun in the latter part of the fifteenth century, came to dominate production on Madeira.

Likewise in Brazil, colonization proper (occupation, settlement, and development) was shaped from the outset by preoccupations mainly political in nature: aiming, through peopling, to retain possessions already being disputed by Dutch, English, and French corsairs. However, advice about this goal presented to King John III (by Diogo de Gouveia, among others) pointed already to the example of the Atlantic islands.[26] With the inception of colonization, the charters granting captaincies visibly aimed at agriculture: the earliest donatories received the specific privilege of building and possessing waterwheels and sugar mills. In this manner, the colonization of Portuguese America was functionally organized from the beginning around production for European markets, and so it developed throughout the sixteenth century.

When the Iberian nations lost their privileged position overseas and colonial competition became general, we witness the same adjustment of colonial expansion to the functional lines of the system. At first the Dutch, English, and French assault on the Castilian Antilles, as we've seen, was directed at establishing beachheads from which to pressure the Spanish colonial system. By the middle of the seventeenth century one notes, there too, a change in course: With the introduction of the sugar economy, the islands of the "American Mediterranean" organized themselves into producers for European markets.

In those areas of the New World reserved to them through priorities of discovery and papal adjustments, the Spaniards were faced with a more densely concentrated population of a higher cultural level than the Portuguese. The previous accumulation of wealth by the inhabitants, along with difficulties of undertaking with them a purely commercial exchange, led the Spanish to a third alternative: conquest, that is, the sacking of accumulated wealth and domination of the indigenous population, along with the direct dismantling of their traditional political structures. Spanish conquest exposed modern colonization's naked coerciveness. Once past that phase, their colonization organized around the mining of silver and gold, the central axis about which everything else revolved: so that even in this case, production for the European market dominated the colonizing process.

Finally, in North America we witness once again the same shift. Colonized since 1607 (Virginia settlement), emigration to these areas had a different connotation. Although the fundamental European expansionary impulses were present, other components interfered in the English version, blurring the results. Even there, emigration to the various American colonies was organized through the mediation of companies that engaged workers for North Atlantic regions, with an eye to colonial profits. But other cases arose through spontaneous emigration of groups persecuted as a result of English political and religious turmoil during the organizational phase of the modern state. The company system, as a rule, did poorly; almost all of them failed. The difficulties of producing products complementary to the metropolitan economy was one factor; others will be examined further on. At the end of the seventeenth century, however, expansion of European tobacco consumption finally opened up for the English colonies south of the Delaware the possibility of meshing with European commercial lines. Virginia, especially, underwent a rapid transformation from a settlement colony, organized on the basis of small and medium-sized properties of diversified production, to an exploitation colony organized into large slaveholding properties producing for the external market.

It was only in the most northerly areas—especially New England, geographically situated within a temperate climate where the possibility of setting up a complementary economy was either reduced or ruled out by the natural setting—that the old configurations of a settlement colony persisted. The establishment of plantations in the south (on the continent and in the islands of the Antilles) that specialized in export goods—and hence short of nutritional and manufactured products—opened up for those northern colonies the possibility of an external market for timber, cereals, manufactures, and so forth. The relative proximity of the two types of structurally divergent colonies thus created an entirely new situation, one particularly favorable to the northern settlement colonies. At the same time, the markets of the metropolis were less interesting to these colonies, for they could but furnish products similar to those available in Europe, and were therefore unable to fit themselves into the role of dependent economies. The diversified subsistence economy directed at internal consumption that characterized these colonies had little capacity for developing a high level of productivity and income until external markets were opened up to them. What is fundamentally important to emphasize, then, is that these markets, when they did open up, were by nature different from the external market common to the other colonies. Within the colonial system as a system the colonies' external market was the metropolitan market. The linkage was made through the regime of the "exclusive," which promoted the exploitation of the colony by the metropolis. In the case of New England, the external market consisted of other colonies: English, French, Dutch, and Spanish. This meant that the relationship that got established was not rooted in the mechanisms of the system; so, the income generated in this relation was not carted off (as was the rule in metropolis-colony relations) but was instead concentrated in the exporting economy itself. This is the fundamental point that permits the understanding of the later development of these settlement colonies, all in all surprising in the frame of the colonial system. They formed an exception, being "colonies" only in nominal political statutes—they were not colonies in a rigorous, structural sense. But bear in mind that it is in relation to the colonial system that they can be understood, even in their anomaly.

On the whole, then, one can now discern the general trend that characterized the setting up of modern colonization within the mechanisms of the colonial system: initial settling, with production for local consumption, followed by interlacing into European commercial lines, and hence into the mechanisms of the European economy. On shifting to production for the external market they would then fully mesh with the larger system since that commercial regime is, as we have seen, the system's nerve center. Thus, the process of colonization adjusts to the thrust of mercantile capitalism's colonial system: the primitive accumulation of capital in European economies was promoted through exploitation of overseas areas.

Not just production, but its rhythm also had to adjust to the system, and, in the end, to the European market, for fluctuations in European demand for overseas goods (*Kolonialwaren*) defined colonial production to a greater or lesser extent. Clearly, alongside the essential production for the European market, another entire sector dependent on it was organized within the colony, which sought to supply internal subsistence needs. But even here we find the colonial system's mechanisms defining the whole and impressing a production rhythm upon it. During periods in which external demand contracted, that is, when European prices for colonial products fell off, plantations tended to shift toward subsistence production because their capacity to import was diminished. When, on the contrary, external demand increased, colonial productive units tended to mobilize all their resources for export production; only then did the possibility of autonomous self-development open up for the colonial subsistence economy. On the whole, then, the exporting sector commanded the entire productive process.

Further, all the structuring of colonial economic activities as well as the societal form for which they serve as a base, was determined within the mercantilist colonial system, that is, within its connections with commercial capitalism. In fact, not just the concentration of productive factors in the fabrication of key products, nor the mere volume and rhythm of their production, but also the actual *mode* of their *production* was defined by the mechanisms of the colonial system. Here we touch on its weakest point. Colonization was organized with the intent of promoting the primitive accumulation of capital within the framework of the European economy, or, in other words, to stimulate bourgeois progress in Western society. This was the deeper meaning that brought together all parts of the system. In the first place, the commercial regime developed within the framework of the metropolitan monopoly; from there, colonial production oriented itself toward those products complementary or indispensable to the central economies; and, finally, production was organized and molded so as to permit the overall functioning of the system. In short, the production of wares for which there was a growing demand in European markets was not enough; it was indispensable that their production be organized in such a manner that their commercialization would serve to provide a stimulus to bourgeois accumulation within the European economies. It was not just a matter of producing for commerce, but rather for a special type of commerce—colonial commerce. It was, once again, that ultimate goal (acceleration of primitive capital accumulation) that commanded the entire process of colonization. This imperative forced the colonial economies to be organized in a mold that would allow the functioning of the colonial system of exploitation, and this, in turn, imposed the adoption of compulsory forms of work, or, taken to its limit, *slavery*.

So Europe was able to contemplate the truly edifying spectacle of the rebirth of slavery precisely at a time when Western civilization was taking decisive steps toward the suppression of compulsory labor and the diffusion of "free," that is, wage labor. While in the sixteenth, seventeenth, and eighteenth centuries Europe made the transition from feudal serfdom to wage labor, which came completely to dominate productive relations with the advent of the Industrial Revolution, overseas, that is, where the world was being Europeanized, the raw, monstrous hulk of slavery reappeared with unprecedented intensity and size. We know for certain that the perplexity created in Christian consciousness by such a situation gave rise, on the one hand, to a vigorous lineage of publicists who, without hesitation, denounced the horrors of modern slavery, and, on the other, to remarkable mental contortions rationalizing slavery and associating it with Christian morality. And Marx used to say that the colonies wind up revealing the secret of capitalist societies. . . .

Let us take a closer look at this point, for it is fundamental to understanding the whole of the system we are analyzing. Slavery was the labor regime preponderating in New World colonization; the slave trade that fed it was one of colonial commerce's most lucrative sectors. If in addition to slavery we add up the various other compulsory, servile and semiservile forms of labor—"encomienda," "mita," "indentured servants"—we find that an extremely narrow strip was left, in the whole of the colonial world, for "free" labor. Old Regime colonization was, then, the ideal universe of unfree labor, the Eldorado that enriched Europe. Explanation of that fact has, at times, bordered on the picturesque. It has been argued, for example, that the Europeans had "resorted" to African labor because population in the homeland with which to people the New World had thinned out. The affirmation handily applies to situations such as obtained between Brazil and Portugal; but if we switch the situations, for example, the French metropolis in the face of the Antilles, the argument does not make sense; there a settlement colony was initiated that only later gave way to slavery. And in certain areas settlement continued to predominate. Anyway, the argument could merely be that Europeans or the European metropolises had at their disposal insufficient demographic contingents with which to people the Americas, and that they then "appealed" to Africa. Nowhere in this argument is there an explanation for why this "appeal" should end up in the enslavement of blacks. What really needs explaining is the regimen of slave labor.

Was it, in any case, really a matter of populating? Within the framework of the colonial system, it was, in essence, a matter of developing new areas so as to promote the primitive accumulation of capital in the metropolises. This naturally involved setting up a productive apparatus, and hence, yes, occupation and settling; but the exploitation of resources for Europe was the essential aim. That is why occupation, that is, geographic expansion, tended toward certain areas (tropical and semitropical), and settlement was organized through the recruitment of workers (whether European, indigenous, or African). The labor regime—the various forms of compulsory labor—still remain to be explained.

But as we have seen, colonial production basically amounted to production for the metropolitan market, that is, mercantile production. Within the market economy, wage labor is the most profitable. The compulsory forms of labor (ancient slavery, and, above all, feudal serfdom) had for their part hinged on premercantile economies (e.g., the closed manorial economy of the Middle Ages). The emergence of mercantile economies (the development of commerce) tended precisely to promote the untying of servile bonds, gradually creating conditions for the expansion of "free" labor—that was the process under way in the early modern period. Mercantilization of production can only become general—dominating all social relations—when the productive force of labor itself becomes a merchandise, that is, when the mercantile economy transforms itself into capitalism proper. To simplify: within this capitalist structure, the productive process begins with an investment of capital (a certain *quantum* of value) in its original form, money, which, on being invested, is transformed into factors of production (productive capital); interaction among these factors generates merchandise, another form of capital (goods), which, when sold in the market, restores capital to its original monetary form supplemented by its surplus value; it is then used to remunerate the various factors (interest, profits, rents, wages) allowing reinvestment at a higher level. Capitalist production is thus steadily amplified, through self-stimulation.

Now it is evident that only wage labor makes possible this type of functioning: if the regime is slaveholding, the cycle binds up, since payment for the labor factor has to be partially made beforehand (in the purchase of slaves), while in the wage labor version remuneration takes place only after consumption of the labor commodity in the actual productive process. The feeding and care of the slave-merchandise stretches out the cycle throughout the slave's lifetime, further jamming the system. In addition, the capitalist system's extraordinary flexibility gets blocked because production cannot adjust itself to fluctuations in demand—for it is impossible to dispense with a labor factor that has been engaged once and for all. Slave labor is thus less lucrative for mercantile production. In that sense, the overseas labor regime prevalent at the very time of the Old Regime was, on the face of it, a breach of good sense. Adam Smith considered such an absurd institution the fruit of slaveholders' pride and love of domination.[27]

Notwithstanding, the slave system dominated the scene of mercantilism's colonial economy. This did not come about through stupidity on the part of colonial businessmen or as a result of flaws in their domineering character. Analysis of the problem simply cannot be limited to a formal logical level.

Examined by itself, the functional aspects of mercantile production preclude the use of slaves for market production. Karl Marx, who framed an analysis of bourgeois society not only from a logical but also from a historical perspective—that is, simultaneously explaining both the mechanics of its functioning and the conditions of its emergence—did not lose sight of the fact that capitalism emerged out of the disintegration of the preexisting feudal-servile and artisan-based (independent producers) structure. Hence, he understood that the development of mercantile relations during the dissolution of the old structure, while enhancing the division of labor and specialization in production, at the same time generated markets thereby permitting the powering of the process. In the most decisive step toward constituting capitalism proper, dissolution of traditional social bonds fostered expansion of the wage labor regime: a process that presupposes, on the one hand, liberation of the worker from all servile obligations, but on the other, at the same time, a dissociation between the producer and his instruments of production, depriving him of any productive factors other than his own labor. In its historical unfolding, the evolution of "free" wage labor involved, on one side, the overcoming of servile bonds (obligations, homage, etc.) and, on the other, a differentiation between the labor of the direct producers and all other factors of production, such as the peasant-serfs' rights over the land, the instruments with which they produced their subsistence, or independent artisan producers' tools. We lack space here to study the lengthy history behind the gradual development of the wage labor regime in Europe. It was through it, however, that the work force emerged at its purest, compelled to trade its labor in the market. If workers had remained linked to the means of production instead of merely renting out their labor power to the owners of those means, they would have made use of those factors to sell their product themselves as autonomous producers, and the capitalist would have had no place in the sun. Whereas, isolated from the other components in the productive process, labor became a commodity like any other. Thus was consolidated the capitalist mode of production, although it was only during the Industrial Revolution that the process of constituting capitalism acquired an irreversible impetus.

To the bourgeois consciousness, of course, what was seen in this lengthy historic process forming the wage regime was the "liberation" of labor from servile injunctions, an ancient barbarism. For, in the capitalist economy, the commercialization of labor hides the new form of exploitation: the generation of surplus value in the hands of the capitalist.

The same Marx—always the implacable analyst of the bourgeois world—precisely because he had carried his study beyond all mystifications was able straightforwardly to affirm that the colonies did not favor conditions allowing the constitution of a "free" labor regime, because in the colonies there was always the possibility of a direct producer transforming himself into an independent producer by appropriating a vacant plot of land. So, while the development of capitalism in modern Europe "liberated" direct producers from medieval serfdom and integrated them into the new productive structure as wage workers thus camouflaging the exploitation of labor, the peripheral colonial economies, erected precisely as levers to foment the growth of capitalism and intimately integrated into its framework, stripped that same exploitation of all pretensions, exposing its darkest, rawest nature. The colonies displayed Europe's viscera.

Eric Williams, in reapplying Marxist analyses to study the genesis of modern slavery, noted quite correctly that the implantation of colonial slavery, far from having been an option (wage worker or slave), was in fact imposed by historico-economic conditions.[28] And here again we face the deeper meaning of colonization and the Old Colonial System's mechanisms, touching on the essential point needed for understanding it. Given the historical conditions attending the colonization of the Americas, the implantation of compulsory forms of labor derived directly from the need of the colonizing entity to have colonies fit in with the mechanisms of the Old Colonial System, that is, to promote the primitive accumulation of capital in the European economy. Otherwise, given the abundance of one productive factor (land), the result would have been the overseas constitution of European settlements developing a subsistence economy directed toward their own consumption, with no effective economic interconnections with the dynamic metropolitan centers. That, however, did not figure into European mercantile capitalism's expansionist impulses as it did not address its needs.

Nothing, in principle, negated the possibility of a colonization in that more restricted sense (of occupation, settlement, and development of new regions). At that point in Western history, however, it was a matter of colonizing for capitalism, that is, in accordance with the system's mechanisms, and those mechanisms imposed compulsory labor. Otherwise, no production for the European market would have taken place as the colonists would have developed an economy turned toward their own consumption. Or, if one were to imagine an export production organized by businessmen who paid their labor wages, those wages would have had to be at a level that would compensate the direct producers for abandoning the alternative of becoming autonomous producers of their own subsistence. In that case, the costs of production would have been such as to preclude colonial exploitation and hence prevent colonization from playing its role in the development of European capitalism. How, then, could the mechanisms of the commercial "exclusive" have been made to function? In short, production for the European market within the frame of colonial commerce, tending to promote the primitive accumulation of capital in European economies, required compulsory forms of labor.

From another angle, colonial production for export—of a volume and at a rhythm defined by European markets and therefore attending to the needs of capitalist development—could only adjust itself to the needs of the colonial system through organization of large-scale production, a move that presupposed ample initial investments. With that requirement, the possibility of production organized on the basis of small autonomous proprietors—who might produce their subsistence, exporting a meager surplus—was also excluded. By analytically examining the impossibility of these alternatives for men living in that era who undertook capitalist colonization, slave (or slave-prone) production reveals itself, as Eric Williams observed, as the almost "natural" choice.

Accordingly, there developed a colonization in the New World centered on the production of staple goods bound for the European market, production firmly rooted in various forms of compulsory labor—at the limit, slavery. Colonial exploitation meant, in the final analysis, the exploitation of slave labor. Thus, the colonists metamorphosed into slaveholders, thereby assuming their destined role in the great world theater. Nor is it any wonder that they yielded to the pleasure of dominating other people—that was merely the misery of the human condition caught in the system's web.

The enslavement of blacks effectively takes us back to the very beginning of overseas expansion. Gomes Eanes de Zurara memorably described the arrival of the first slaves in Christian Europe.[29] But these earliest supplies of slave merchandise were destined for European "consumption" during a precolonial commercial expansion, and that introduction of slave labor into the midst of an expanding capitalist-mercantile economy was not widespread. Only because of the conditions in the colonies that we have just examined did slavery find fertile soil for development in the overseas colonial world. In the Atlantic islands—site of the first modern colonization—the initial settlement, with its diversified, self-sufficient economy, gradually shifted toward specialized production for the metropolitan market, bringing with it a new labor regime. The next step was the introduction of African slavery: "[sugar] cultivation expanded into a new world; it prospered, while Africa was despoiled of her wilderness children so the civilized might dine cheaply."[30]

With agro-industry transplanted to Brazil during a phase in which large-scale consumption predominated and prices were rising anew, the natives were at first compelled to perform the arduous work of cane cultivation and sugar refining. Expansion of production utilizing an ever larger enslaved labor force soon gave rise to the Negro slave trade to the New World. "It is unquestionable," said the historian Lúcio de Azevedo, "that we owe to sugar the development of slavery at the core of modern civilization"[31]—which is probably an exaggeratedly synthetic way of putting it: All of the complex interweaving of the colonial system is connoted by the word "sugar." But it was on this slaveholding basis that the colonization of Portuguese America was developed, and colonial society continued to be molded on that same basis. The Jesuit Manoel da Nóbrega noted, in Brazilian colonization's early days, that "the men who come here find no way other than to live off of the work of slaves."[32]

The introduction of African slaves has been explained, from one side, curiously, by the "inadaptation" of the Indians to tillage, and from another by citing Jesuit opposition to Indian enslavement. There is no doubt that Jesuit sermonizing weighed in the aborigines' defense, though we may note in passing that this failed to safeguard them entirely: whenever there was a shortage of Africans (due to difficulties in navigating the Atlantic or due to colonial competition, for example) the Portuguese immediately turned to the compulsion of natives. It is also true that the blacks could not rely on that same Jesuit defense. The arguments justifying such a discrepancy were truly edifying, but it does not behoove us here to delve into theological questions. What appears indisputable to us is that not only were the natives made use of at certain times, especially during the initial phase, but that one certainly cannot allege any greater or lesser "aptitude" to slave labor (for that is what the argument amounts to). What may have mattered was the demographic thinning out of the indigenous population, along with difficulties in their supply, transport, and so on.

But in the preference for Africans,[33] there reappears, we believe, yet another time, the workings of the mercantilist system of colonization. This system unfolds, we repeat as often as necessary, within a system of relations tending to promote primitive accumulation in the metropolis; the trafficking in Africans, that is, the supplying of the colonies with slaves, opened a new and important sector of colonial commerce, whereas the capture of Indians was an internal colonial business. Thus, the commercial profits deriving from the supply of Indians were kept in the colony by the colonists. The accumulation generated through the trade in Africans, however, flowed into the metropolis and was realized by metropolitan merchants engaged in furnishing this "merchandise." This may well be the inner secret of the better "adaptation" of the Negro to tillage—to slavery. Paradoxically, it is by commencing with the slave trade that one is able to understand colonial slavery, and not the other way around.

In the English, French, Dutch, and Spanish colonies, instances of the phenomenon exhibit regional variations—we lack space here to analyze the manifestations of the phenomenon in all its minutiae—but the background remains the same. All relied on varying forms of compulsory labor, servile or semiservile, especially slavery, that dominated overseas production and defined colonial society.

The Crisis of Mercantilist Colonialism

Such were the parts of the system, and its functional mechanisms; we now have at our disposal the elements with which to analyze its crisis.

For if we think in terms of *the system's* crisis, then it is from its own functioning that that crisis must emerge, and not from exogenous factors. In short, as it evolved, the Old Regime's colonial system simultaneously fostered elements that led to its own eclipse.[34]

And, in fact, while colonization in the mercantilist era developed by promoting primitive capital accumulation for the European central economies, toward that end (i.e., in order for colonial exploitation to continue to function) it went about engendering within the overseas world the universe of the seigneurial slaveholding society. Yet the interrelations and values of this society became increasingly opposed to those held by the rising European bourgeois society.

Let us linger, however, for just another moment, on the implications of slavery for colonial societies and economies. To begin with, within the sphere of production one immediately distinguishes two basic sectors. The first, an export sector, organized into large productive units functioning on the basis of slave labor and centered on the production of goods for European consumption, was the essential sector that addressed the very reason for capitalist production. The other, subordinate to and dependent on the first, was a subsistence economy to attend the needs of local consumers with goods not imported from the metropolis. It fell to small proprietors and independent laborers and was organized so as to allow the functioning of the first. The dynamic of the whole of the colonial economy was determined by the exporting sector. In some circumstances and within given areas, the subsistence sector could, as in the cattle industry, loom somewhat large, organizing into extensive properties, or, as in other cases, incorporating the slaveholding regime. But the global dynamic always depended on the external flux, the ultimate primary center being European capitalism. It meant, in every sense of the term, a dependent economy—with the principal sector directly dependent, and the secondary sector indirectly so.

Second, at the level of socioeconomic relations, the slaveholding structure greatly increased the concentration of wealth in the hands of the slavemasters who owned the enterprises that produced goods for colonial trade. The direct producer, reduced to the status of a mere instrument of labor—*instrumentum vocale*—that is, a person transformed into a thing as a slave, did not, by definition, possess an income of his own; the earnings thus concentrate in the seigneurial stratum. At this point, we meet up with the element we lacked for our understanding of the system's mechanisms: This concentration of income was precisely what was necessary in colonial society, permitting it to function and finally hinging together the various parts of the mechanism. Note well: the overall income generated by the peripheral economies was only finally realized in the European central economies' markets; thus, the greater part of it was transferred, through the colonial commercial mechanisms analyzed above, to the metropolises, or better, to those bourgeois groups connected to overseas transactions. But it happens that the concentration of the remaining minor parcel retained within the colony among a small seigneurial stratum was what made the continuous functioning of colonial exploitation possible. It was this concentration of income that made it possible, despite the transfer of the larger share to the European bourgeoisie, that the seigneurial colonists were still able to maintain continuity in the productive process to the point of affording themselves luxurious standards of living. By the same token, still within the mechanisms of the system, the very colonists themselves had resources with which to import products of the European economy. The most significant part of their income was created through exports and consumed in imports, transactions made within the colonial commercial regime, and which transferred the profits to the metropolis, so that, viewed as a whole, colonial society was exploited by the metropolitan bourgeoisie. But within that selfsame colonial society, the colonial seigneurial class found itself in a privileged position, and this fact made possible the meshing of all the various parts of the system. And slavery, which is the other side of the same coin, reappears as its essential element. Now, viewed from another angle, colonial exploitation means the exploitation of slave labor.

The implications derived from the mode assumed by colonial production do not end there.[35] Slave-based production for the European market, that is, the *mercantile-slave* system, was characterized by a shortage of local capital (linked to the exploitation of the colony by the metropolis) and an abundance of land (we've already noted the structural connections between availability of land and introduction of slavery).

Furthermore, the very structure of slaveholding hindered investment in technology. The slave, as a slave, has to be kept at subhuman cultural levels so as not to awaken to the true nature of his human condition—that is an indispensable part of slaveholder's domination. It follows that the slave (as slave) is not apt to assimilate more advanced technological processes. In some situations the seigneurial colonists came even to oppose, to popular amazement, the teaching of the catechism to slaves (which, after all, had been the very argument used to justify their being drawn from Africa), for it posed dangers: learning a common language would enable communication among various African groups. We must remember that, contrary to the common supposition, it is an illusion that a slaveholding society enjoys stability. In fact, flight and rebellion were quite common, and the ubiquitous whipping posts could not, even from a distance, be mistaken for decorative objects. Let us not, however, stray too far from our reflections: capital was not available, and the slaveholding structure did not favor technical progress.

The result: the colonial economy suffered from low productivity. As a consequence, as noted by Celso Furtado, it grew extensively, that is, through aggregation of new units composed of the same factors.[36] But still, since it did not reinvest on an increasing scale, but simply replaced and aggregated—it depleted natural resources. The mercantile-slave economy was a predatory economy. Again we meet up with the primary purpose of colonization: the unfolding of European commercial expansion. Just as New World colonization began as a purely commercial activity involving natural products (dye-wood, furs), just so, although with the emergence of colonial production the system took on extraordinary complexity, it kept on depredating natural resources. In this sense, then, colonial expansion faced natural limits: the draining off of resources dilapidated by the colonial mode of production. Since, meanwhile, this process developed within a larger context and not just in a purely economic one in its strictest sense, long before any of those limits had been reached, tensions of every type began to appear. Thus, we begin to uncover the contradictions within the system.

The slave structure of the colonial economy and society effectively implied, albeit indirectly, an eventual limitation on the market economy's growth. The contradiction thus points up the very nature of colonial production: at once both mercantile and slaveholding, that is, producing goods for European capitalism through the labor of slaves. Those two defining components of the colonial economy coexisted with difficulty within the same context, generating rising tensions. On one hand, slavery brought about a low level of productivity and, hence, only a limited profitability in colonial production (as distinct from its commerce). But since it lacked the means for minimizing costs through technical progress, the seigneurial-enterprising class necessarily had to try to minimize the cost of maintaining the slave-labor force. To that end, it tried to get from the labor of slaves at least a portion of their own subsistence within the estate itself. Thus, there was inserted into the core of a basically mercantile economy, an entire segment of subsistence production whose activities took place along the edge of and almost outside the market economy. Added to this was that part of the payment for the labor factor went outside the producing area (the payment for slaves made to European merchants), while the other part (for the maintenance of the slaves) took place through subsistence production, leaving no place for mercantile operations within the colony, at least not on a large scale. Therefore, neither of the two sectors into which the remuneration of labor was divided within the colonial economy created an internal demand that could have stimulated economic development. To sum up: the mercantile colonial slave economy necessarily had a

Brazil in the Old Colonial System

very small internal market.

Within the whole of the system, this meant that the colonial economy became ever more dependent on the metropolitan economy. Given the narrowness of the internal market, it had no means of stimulating itself, thereby being left subject to the impulses of the dominant dynamic center, that is, of European commercial capitalism. In that sense, the phenomenon adjusted itself to the system and there were no contradictions.

But let us examine it from another angle. On the one hand, European expansion had meant, basically, commercial expansion, the opening of advantageous new markets—and colonization, as we have seen, had meant an extension of the market economy. Looked at internally, however, colonial economies revealed themselves as a mercantile-slave mode of production that limited the constitution of their internal market. There was a substantial layer of the population (the direct producers) whose consumption, in large part, was effected outside of mercantile transactions. It was truly an expansion of the market economy, but it carried within it structural limitations to that expansion. What followed from this was extremely important. In the typical colonial (mercantile-slave) economy, or, more precisely, within colonial society, the universe of mercantile relations affected only the upper social strata of slaveholding colonists.[37] They only imported merchandise from the central economies for their own consumption: food or manufactured goods for personal consumption, implements for productive consumption.

To be sure, the reality was somewhat more complex. Colonization involved other activities (administrative, military, religious, etc.) that amplified to a certain extent the segment of colonial society connected to the mercantile economy. The actual functioning of colonial production demanded social categories other than the master-slave binomial. The sugar industry, for example, required an entire range of skilled workers, technicians, suppliers, and so on. Commerce itself relied on intermediary personnel, clerks and accountants, for instance. All of this resulted, within the colonies, in the formation of the earliest urban settlements that enlarged the sector involved in the market economy.

But note that all of the components of colonial society that we have been indicating (functionaries, administrators, clerics, military) were basically secondary categories of colonial society to the extent that their presence in the overseas world followed from the slaveholding economy and from production for European capitalism. Colonization was undertaken to produce for the metropolis, and colonization ended up involving other components. The other social categories, however, depended on the master-slave matrix the same way as the subsistence sector depended on the export sector. At root, therefore, within the colonial environment, everything depended on the seigneurial class, and the mercantile economy within the colonies expanded as a function of it. The fundamental mechanism remained. The universe of mercantile relations was a function of the masters and, shall we say, their entourages. The mass of direct producers (the slaves) lived outside mercantile relations, and this hindered the constitution of an internal market.

On the whole, such a configuration of the colonial world responded adequately to the functioning of the system as long as the central economies only developed at the level of the primitive accumulation of capital and production only expanded through handcrafts or even manufacture in the true sense of the word. However, once this stage was surpassed, and the mechanization of production in Europe reached toward the Industrial Revolution, invigorating productivity at a rapid pace and in an intense way, it led to a growth of capitalist production at a volume and rhythm that came to demand much more ample overseas consuming centers, consumption not only by upper societal strata, but by the colonial society as a whole. And this required the generalization of mercantile relations in the colonies themselves. Then the system was compromised and entered a crisis.

Now, in promoting the primitive accumulation of capital in the European central economies the colonial system had acted, as we have seen, as a fundamental instrument for promoting the transition into industrial capitalism. (Of course, it was not the only element: one must consider the internal factors related to the development of capitalism in Europe.) In organizing themselves within the framework of the colonial system, the peripheral economies developed their production along lines tending to complement the central economy, supplying those products that it needed and providing raw materials first for European manufactures and later for the factory system. They were thus configured into authentic complementary economies, tending to bestow on the metropolises autonomous economic conditions in competition with other mercantilist nations. Note the importance of this mechanism, in an era during which the practices of mercantilist policy became general among the various European states. The colonial markets were those where, by definition, mercantilist rules could be exercised: hence, the truly furious disputes over the conquest of those exceptional markets.

Along these same lines, there had developed in the metropolises a colonial policy aimed at obstructing manufacturing production within the colonies. They sought in this way to reserve the colonial market for the manufactures of the homeland. Given the social and economic structure into which the typical colony was organized, or at least the ones that were perfectly conformed to the system, the possibility of developing manufactures was basically reduced anyway. The prohibitionist policy encountered weak resistance due to the poor conditions for the growth of manufactures in the colonial world. Accordingly, expansion of overseas colonization efforts effectively came to involve only a gradual enlargement of the manufactured goods consumer market.

Only in the settlement colonies such as New England were such conditions favorable, but New England was regarded in mercantilist circles as "the most prejudicial plantation of this kingdom" (Josiah Child).[38] Likewise, the results of the prohibitionist policies were successful or not according to whether they applied to colonies with a greater or lesser degree of adjustment to the system. The settlement colonies had been constituted, as we have seen, within the New World's temperate zone, a region not targeted for modern European colonization in its early phase precisely because of the impossibility of organizing in that location a type of production able to satisfy the European market's demands. So, during the seventeenth century, English emigrants fleeing the political and religious tensions of the homeland had migrated into the region in an effort to rebuild their way of life in the New World. The settlement colonies were thus formed bordering the system, and the struggle for independence and the establishment of the United States were impelled by metropolitan efforts to make these colonies conform to the larger system. Such events signaled the beginning crisis of the Old Regime.

Adding up the elements analyzed so far, that is, the colonial system's functional mechanisms, we can now explain its position within the rise of capitalism. The colonization of the New World in early modern times, that is, the exploitation of overseas colonies organized along the lines of the Old Colonial System, formed a powerful instrument for accelerating primitive accumulation within the context of European mercantile capitalism. It comprised a process for transferring income from the colonies to the metropolises, or more precisely, from the periphery to the dynamic centers of the European economy. Such income tended to concentrate in the entrepreneurial strata connected with colonial commerce. By being themselves complementary economies, that is to say, economic buttresses for the metropolises, these elements of the Old Colonial System heavily contributed to the development of European national economies; development that, at the time, consisted in expanding mercantile capitalism, thereby hastening capital accumulation.

If we now recall what we indicated earlier regarding commercial capitalism as an intermediate phase between the disintegration of feudalism and the Industrial Revolution, the mercantile colonial system reveals itself as operating on the two basic prerequisites for the transition to industrial capitalism. Overseas colonial exploitation promoted, on the one hand, primitive capital accumulation on the part of the enterprising stratum in Europe; and, on the other, it enlarged the consumer market for manufactured goods. It simultaneously augmented the factory system's productive capacity and increased the demand for manufactured goods. The prerequisites for the Industrial Revolution—the central historic process in the emergence of capitalism—were thus created.

Now we reach the heart of the system's dynamic: in functioning as a whole, it simultaneously brought on the conditions for the onset of its crisis and its denouement. Even before the system's alternatives had been exhausted, that is, prior to reaching the limits inherent in colonial exploitation, the tensions generated by these fundamental mechanisms began to impose new adjustments, alterations, and changes that then compromised the colonial system. In other words, it was not necessary for industrial capitalism to achieve the highest possible levels of development and expansion before the colonial system—slaveholding, colonialist—entered a crisis. As a matter of fact, the first impulse was enough. The first steps into the Industrial Revolution were sufficient to get it under way.

The furious overseas competition among the mercantilist powers was inherent in the very logic of the system of colonial exploitation. Such competition could perforce only result in the hegemony of one of these powers. Nor was it any coincidence that England was both the power that won the race of colonial competition and the nation that took the first steps toward modern industrialism. One should not overlook the internal factors affecting its economic growth on the way to industrialization, but it is clear that its colonial supremacy allowed England to transfer within its borders—to a greater extent than the other powers— the stimuli forthcoming from the colonial system. The consolidation of British preponderance and the opening of the Industrial Revolution converged about the 1760s.

During the critical period between 1763 (end of the Seven Years War) and 1776 (independence of the United States), those problems were acutely felt. Having overcome the rivalry with France, Great Britain sought, on the one hand, to reinforce its own metropolitan "exclusive" (in an effort to bring the New England colonies into conformity with mercantilist policies), and, on the other, to widen commercial inroads into the Hispanic colonies, whether legally through their metropolises or through smuggling. All of this sprang from Britain's political supremacy and industrial development. Yet, just as the system was fully worked out, it engendered tensions of all kinds. The more the process developed, the less England could withstand the "independent" trade of its colonies. At the same time, smuggling with Iberian colonies became itself an insufficient outlet for the flow of its industrial production. Further, the monopoly that the Antillian English plantations enjoyed in the English metropolitan market (the other side of the Colonial Pact) became increasingly onerous to the metropolis: it was as though the Colonial Pact were inverting itself, benefiting the colony rather than the metropolis.

Within this framework of acute tensions, this complex of multiple interests, the equilibrium became increasingly precarious. It broke down entirely with the independence of the United States. The emergence of the new republic in effect carried implications that greatly transcended the merely political event. It was the first time a colony had ever become independent. The tensions, the competition, the appropriation by one power of colonies belonging to others, all had been adjustments within the system. What obviously could not be contained within the system was the breakdown of the Colonial Pact itself. To the extent that structural tensions worsened, causing divergent interests to surface, the colonial world came to exist in a state of constant excitation. Criticisms of the Old Regime were heard also in the colonies and found a highly receptive audience. With the independence of the United States, however, what had once been a possibility came to be reality. The political innovations woven into the republican form adopted by the new state further highlighted the significance of the change, marking the beginning of the crisis not just of the colonial system, but of the entire Old Regime.

Notes

[Ed.: This chapter was translated from Fernando A. Novais, *Estrutura e dinâmica do Antigo Sistema Colonial*. That text first appeared in his "Portugal e Brasil na crise do Antigo Sistema Colonial," PhD diss., chap. 2, and was subsequently included (with some modifications) in his book by the same title. I have shortened the text in a few places where it did not alter the argument.]

1. *Les migrations des peuples*, 11–16ff. [Ed.: Aside from explanatory notes, those that follow are limited to those that give the source of quotations in the text or refer to authors actually mentioned in the text. The full bibliographical basis of this chapter can be found in the original as well as in *Portugal e Brasil*.]

2. "Colleção das leys e ordens que prohibem-os navios estrangeiros, assim os de guerra, como os mercantes, nos portos do Brazil," Ms., Arquivo Histórico Ultramarino (Lisbon), cód. 1.193; also copy at Biblioteca Nacional, Rio de Janeiro, Ms. 7, 1, 6.

3. *Britain's Commercial Interest Explained* (1747) as quoted by Sée, *As origens do capitalismo moderno*, 136. See also the expressive example of the mercantilist concept from the pen of the Marquis of Pombal when addressing the French ambassador in 1776: "the overseas colonies having been established *precisely for their utility to the metropolises to which they belonged*, from

thence were derived the infallible laws, universally observed in the practice of all nations," Pombal to Marquis de Blosset, January 31, 1776, quoted in visconde de Santarem, *Quadro elementar das relações*, VIII, 151–155 (our italics).

4. *La época mercantilista*, 17–29.

5. *Studies in the Development of Capitalism*, 37ff.

6. Stark, *Historia de la economía*, 20–26. Even Hecksher (*La época mercantilista*) may not have been aware of something that did not escape the attention of Lord Keynes, who noted that at a time when governments had practically no way to manipulate interest rates, the best way to keep them low and thus foster productive investments was the abundance of currency: Keynes, *Teoria geral do emprego*, 319–350.

7. [Ed.: In 1979 Novais added at this point the following note:] Without getting into an exhaustive analysis of the transition from feudalism to capitalism, which would be beyond the limits of this chapter, we have indicated in the text the most important connections among the various levels and sectors of historical reality in the modern era. That should be our principal aim at this point. In this regard Immanuel Wallerstein's *The Modern World System* (New York, 1974) is most suggestive.

8. *La Mediterranée et le monde méditerraneén*, 619.

9. For example, "The best indication of her [Brazil's] wealth is that when men return from India to Portugal they bring along all they own because none there have real estate (or if any do their holdings are slight), and all their fortune consists of physical things with which they embark and with the sale price these command in Portugal they assure themselves a life-income or buy houses. Whereas the residents of Brazil have all their wealth in real property, which cannot be taken to Portugal, and when any depart they must leave it in the land" [Brandão?], *Dialogos das grandezas do Brasil*, 79.

10. Prado Júnior, *Formação do Brasil contemporâneo*, 4th ed., 5–26, 113–123.

11. Marx, *El Capital*, I, 801 and esp. 840–851. We lack space in this chapter for a more extensive discussion of primitive accumulation.

12. Weber, *Wirtschaftsgeschichte*, 3d ed., 256–259.

13. Leroy-Beaulieu, *De la colonisation chez les peuples modernes*, 533ff.; or, in the 5th ed., II, 563ff. He based himself on the classification of Roscher and Jannasch, *Kolonien, Kolonialpolitik und Auswanderung*, 3d ed., 2–32.

14. Soon after the rounding of Cape Bojador—a decisive event in the history of European expansion—the Portuguese king forbade navigation to the discovered lands without the authorization of Prince Henry: Carta Regia of 22 October 1443 in Godinho, ed., *Documentos sobre a expansão portuguesa*, I, 142.

15. The phrase is from Dias, *O capitalismo monárquico português*, II, 355ff.

16. Quoted by Holanda and Pantaleão, "Francêses, Holândeses," tomo 1, vol. 1, 165–166.

17. "The notable progress made by this city [Antwerp], so rich and so famous, began around the years 1503 and 1504, when the Portuguese, having earlier occupied Callicut by dint of marvelous and stupendous navigation and having made an accord with that country's king, began bringing spices and drugs from the Indies to Portugal and thence to this city's markets": Ludovico Guicciardini, *Discrittioni di tutte Paesi Bassi* (1567) quoted in Hauser and Renaudet, *Les débuts de l'âge moderne*, 61–62.

18. [Ed.: 1 *arrôba* = 15 kg.]
19. In Santarém, *Memórias e alguns documentos*, 65-66, 222-224.
20. Furtado, *Formação econômica do Brasil*, 1st ed., 20.
21. As can be seen in the table he presents in *Le Portugal et l'Atlantique*, 256, the price of sugar in Brazil remained at 800 *réis* per *arrôba* from 1570 to 1610, whereas in the same period in Lisbon it ran up from 1,400 to 2,020 *réis*. In 1614 prices in these two markets converged (1,000 *réis*) only to diverge again: by 1650 it was 700 *réis* in Brazil and 3,800 in Lisbon.
22. Book V, Titles CVII and CVIII of the *Codigo Philippino*, edited by Almeida, 1253-1259.
23. Guillermo Céspedes del Castillo, "La sociedad colonial americana en los siglos XVI y XVII," in Vicens-Vives, ed., *História social y económica de España y América*, III, 479.
24. Quoted in Clough and Cole, *Economic History of Europe*, 347.
25. For example, Mauro, *Nova história e Novo Mundo*, 61-64.
26. Gouveia quoted by Azevedo, *Epocas de Portugal econômico*, 233-235.
27. *An Inquiry into the Nature and Causes of the Wealth of Nations*, 364-366.
28. *Capitalism and Slavery*.
29. Zurara, *Crônica dos feitos da Guiné*, chap. 24, 122-123.
30. Azevedo, *Epocas de Portugal econômico*, 228.
31. Ibid.
32. *Cartas jesuíticas*, vol. 1, 110.
33. Goulart estimates their number to the end of the eighteenth century at 2,200,000: *A escravidão africana*, 217.
34. Contradictory development appears inherent in various stages of capitalism's colonial exploration; see, for the nineteenth century, Marx's analyses of British domination in India: Marx and Engels, "Sobre el colonialismo."
35. [Ed.: In 1979 Novais added at this point the following note:] To further the analysis of the colonial system's crisis it does not seem necessary to explore in depth the issue of "the colonial mode of production," although our position on this theme is implicit. The works of Ciro F. S. Cardoso are clearly those that have taken this conceptualization the furthest. See his "Severo Martínez Peláez y el carater del régimen colonial" and "Sobre los modos de producción coloniales de América." These two works are also published in *América colonial: ensaios*, ed. Santiago. As is natural, given the complexity of the problem, we do not entirely agree with some of Cardoso's views.
36. *Formação econômica do Brasil*, 66-69.
37. To the extraordinary complexity of the "colonial economy" engendered by the structures of the Old Colonial System is connected the peculiarity of the social formation that it supports. See the reflections of Florestan Fernandes characterizing Brazilian social formation: *Sociedade de classes*, 9-90.
38. *A New Discourse of Trade* (1669) quoted in Clark, *History of Manufactures in the United States*, I, 4.

2. The Patrimonial Dynamic in Colonial Brazil

John R. Hall

Ever since the appearance of Andre Gunder Frank's *Capitalism and Underdevelopment in Latin America*, theories about world systems and dependency have been the subjects of strong debate. Whatever their merits as formal theories, they helped reinvigorate historical sociology.[1] During the same period, historical studies of Brazil experienced a seemingly exponential growth.[2] Much of the new work on Brazil deals, at least implicitly, with questions posed in dependency and world-system theories about world inequality. But critical reaction to the two theories has been substantial enough to give pause to ask: what should be the working relation between these theories and historical research?

The answers do not come easily. Under conditions of epistemological and methodological controversy, practicing historians might take some comfort if they could invoke a "universal history."[3] Such an overall theory of historical development would solve the problem of "selection"—how to choose and order data from manifold possibilities.[4] World-system theory might seem a likely ordering device, but there are good theoretical and substantive bases to conclude that it is not. In this essay, I sketch world-system theory and its limitations as a framework for universal history. As an alternative, I propose a neo-Weberian approach, using substantive models and theories to analyze historical development without recourse to any universal theory of historical change. To illustrate the neo-Weberian approach, and to explore historical dynamics that cannot be reduced to a holistic world-system framework, I address the problem of patrimonialism in the history of colonial Brazil.

World-System Theory and the Weberian Approach

The world-system approach may be understood in one of two ways: either it establishes a particular *level* of analysis, but without privileging any particular *explanation* at that level, or, alternatively, it offers a formal and comprehensive *theory* intended to explain world-historical development. The former alternative simply posits the geopolitical economy as a central subject for historical inquiry. In terms of this basic recognition, world-system analysis would not differ significantly from the work of Marx, Lenin, Weber, or Braudel, all of whom recognized and discussed historical development in such terms. A number of contemporary analysts have adopted this theoretically eclectic world-systems approach, and much of the good research has not been bound by any formal theoretical model.[5]

It is clear, however, that major proponents of world-systems analysis have intended it as much more. For Immanuel Wallerstein, the "world system perspective" encompasses dependency theory and other critiques of the developmentalist approach. The overall perspective is based on the assumption that there is one capitalist world economy, which Wallerstein characterizes as having a single international divison of labor, with relatively highly developed nation-states at its *core*, relatively underdeveloped societies at the *periphery*, and a *semiperiphery* of buffer nation-states undergoing decline from the core or development out of the periphery. Theoretically, relationships between societies in different positions are to be explained primarily on the basis of their patterns of trade with one another; thus, the famous dependency formulation that the more powerful countries have developed poorer countries along lines of underdevelopment. In formal terms, these precepts establish a model in which the economic roles of regions incorporated into the world economy through trade are determined by their functional relations to the world process of capital accumulation. As Wallerstein put it, there is a need to conceptualize "one capitalist economic system with different *sectors* performing different functions." It is this version of the world-system approach that has the trappings of a universal history, for events are ordered by an assumption of holism, in which the totality defines the nature of the parts, as well as their relationships to the totality over time. History unfolds as the growth and reordering of the world system.[6]

Much of the critical reaction to world-system theory challenges the assumption of holism. Marxists have dismissed the "necessitism" (i.e., necessary determinism) and "reductionism" of a holistic assumption. Various critics have argued that the emphases on trade and capital accumulation give short shrift to class formations and class struggle. My own critique has been that the assumption of holism promotes a misleading account of the emergence of the capitalist world economy: the holistic model submerges key continuities between ancient empires and the world economy, and it fails to reckon the independent salience of patrimonial organization.[7] Whatever the resolution of these controversies, their very existence draws into question the capacity of the world-system approach to serve as a master theory of historical change.

Given the significance of questions posed in world-systems analysis, how to proceed? One alternative is to abandon the search for an encompassing theory of history in favor of a neo-Weberian approach that encourages analytic historiography, but without privileging any particular explanation in advance. This approach comes closer to organizing historical research in terms appropriate to the present play of conflicting explanations.

In scholarship on Latin America such conflicts run deep. Although Latin American historians have recognized for some time that dependency theory offers a fruitful course of inquiry, they have been slow to embrace the world-system approach.[8] But they have not often pursued the potential of Max Weber's strategy either; as Stuart Schwartz observed in the early 1970s, studies drawing on Weber's work had tended toward mechanistic application of "ideal types" as static models. In an era when social theorist Talcott Parsons claimed to supersede Weber's insights, historians were blocked from fully appreciating the more complex Weber who had offered a comparative and configurational history of long-term social change.[9]

By now, understandings of Weber have advanced, and the potential of his strategy is worth considering. What is the structure of Weber's approach? Briefly, Weber consolidated a working relation between historical analysis and sociological categories that depends on *Verstehen*, or meaningful interpretation of social action. For others, such a thoroughgoing "methodological individualism" would have been a difficult foundation on which to build structural analysis of large social complexes, and one that might easily lapse into purely ideographic explanation. Weber avoided these pitfalls largely by using basic sociological concepts and general sociohistorical models (often called ideal types) to formulate and assess substantive theoretical explanations tied to historical research.

The process of ideal type concept formation consolidates sociologically consistent features of social actions and situations in ways that are unlikely to be found in reality in their pure form, yet offer close parallels to historical phenomena. Thus, no actual organization exactly fits the legal-rational bureaucracy of the type Weber described, but the type is useful as a basis for exploring the dynamics and charting the changes of actual organizations. Such sociohistorical types lack the rich detail of direct historical description, but Weber argued that they compensate for this limitation by offering a "greater precision" in the analysis of social action, organizations, and institutions.[10] In *Economy and Society*, Weber offered an encyclopedic array of concepts about action, types of productive organization, types of economic exchange, classes, status groups, parties, political action and alliances, and structures of organization. Paralleling this typification of action and organization, on the level of knowledge, ideology, and culture, he identified alternative legitimations of power, ethics of action, and theories of salvation. The result is a battery of concepts that are at once based in constructs of meaningful action and, at the same time, capable of being applied to the discussion of "structural" phenomena such as political and economic organization. Thus, the controversy between idealist and structuralist readings of Weber seems a red herring, for Weber's concepts are not dependent solely on idiosyncratic individual motives for their salience, yet retain subjective orientation as a basis of conceptualization.[11] Nor does Weber's approach presume either a particular "unit" of analysis (e.g., a "society," "social formation," "world system," or indeed any unitary reality) or any theoretically necessary linkages among concepts. In lieu of positing causal linkages at the theoretical level, the burden of explanation is shifted to the historical world, where men and women enact overlapping patterned dramas and struggles of social life.

With arrays of typological models as building blocks, research may move in several basic directions. First, it may "precipitate out" aspects of a phenomenon that may be explained in terms of one or more models. The "residual" that remains unexplained then must be subjected to more detailed analysis of the unique situation. Second, the relevance of a given model of action or organization may be traced historically, and the points identified at which there is a shift, transformation, accommodation, or assimilation. For example, it would be possible to map the emergence of legal-rational bureaucratic organization in absolutist states. Here, again, pinpointing the circumstances of change offers the opportunity for specific historical explanation. These sorts of studies amount to what Guenther Roth has termed "situational" and "developmental" histories. Third, a historical phenomenon may be compared to others within the range of a particular sociohistorical model's applicability, or to other cases of another type, to explore patterns and the factors that condition development along one or another line. Finally, a phenomenon can be described in terms of a battery of models; then the various elements may be traced both to their historical origins and to the causal sequences that gave rise to the configuration.[12]

Max Weber is best known in scholarship for the particular ways that he addressed the rise of Western capitalism in his study of the Protestant ethic and his comparative studies in "religion." Politically he was preoccupied with nationalistic concerns about the instabilities that racked Germany. Yet the approach just described can be applied to scholarly problems other than Weber's, and by people who do not share his political values. Thus, in the post-Parsonian era, some sociologists have identified themselves as "left Weberians" and a wider set has claimed a "neo-Weberian" or simply Weberian approach.[13] By their efforts, the approach has come to represent more than the corpus of Weber's work alone. It stands as a distinct alternative to contemporary evolutionary, stage, and systems theories of historical development. The neo-Weberian approach counters theories based in holism and necessitism, which force events into the matrix of a universal history founded on some prime mover—materialist, idealist, or otherwise. Yet Weber's strategy does not retreat from a general theory of history into empiricism, as some critics claim. He chose a middle course, recognizing that underlying, sociologically understandable dynamics structure the "surface" events of history, yet without assuming that these dynamics are linked transcendentally by some essence.

The Weberian Approach, Brazilian History, and Patrimonialism

Weber addressed issues of colonialism, world economy, and geopolitics more with an eye to understanding their significance for industrial capitalism than for their effects on the periphery. But he developed a comprehensive account of agricultural organization, precapitalist forms of social organization, commerce, political and military domination, colonialism, and industrialization, and these categories are well suited to the purposes of colonial and postcolonial historiography, both in general and in Brazil.[14]

In colonial Brazil, the history of uneven development turns on (1) the relationships among the ruling regime and its patrimonial offices and bureaucratic and military apparatus, (2) export-oriented agricultural classes of landowners and a merchant class (and, at the end of the colonial era, a nascent manufacturing bourgeoisie), all in their relationships to (3) foreign powers and foreign capitalists. Other major strata of Brazilian society—slaves, peasants, the nascent proletariat, and urban masses—have until very recently been largely the objects of manipulation, providing the labor to make history not under conditions of their own choosing. The most critical questions concern the relationships among the key players of Brazilian history. Are we to understand actions of the ruling regime simply as dictated by the interests of world capital and its accumulation? How are we to account for the relative inability of Brazil's urban merchant class to influence state policy? These sorts of questions cannot be answered by theoretical fiat, nor simply by cataloging the findings of Brazilian historiography to date. They require consideration of these findings in relation to one or another explanatory model.

How might Weber's sociological matrix illuminate the colonial history of Brazil? In many ways, Weber's secular theories resemble those used by others, from Marx to Wallerstein, to explain socioeconomic development.[15] Yet at least one of Weber's ideal types does not fit easily within either Marx's or Wallerstein's formulation. Empirically, when analyzing barriers to capitalist development, Weber drew heavily on a sociohistorical model of patrimonialism. As a way of demonstrating the potential of a neo-Weberian analysis, I will pursue the following claim: Patrimonialism is the axis around which much of Brazil's history turns. It then will be possible to ask whether historical dynamics of patrimonialism in Brazil are at odds with the holistic assumptions of world-system theory. Other writers, for example, Riordan Roett, already have argued the patrimonial character of Brazilian society in a general way, or, like Uricoechea, they have shown the significance of patrimonial organization for a particular social stratum. The pioneering formulation in the 1950s was offered by Raymundo Faoro, in *Os donos do poder*.[16] My intention here is to build on this body of work by constructing a developmental account of patrimonialism and the political economy of colonial Brazil.

While there is much to be said for economic histories that show Brazil's connections to the world economy and the political realities that turn on those connections, the patrimonial features of Brazilian society represent something of a missing link, one that connects Brazil on the one hand to the ancient patrimonial regimes that preceded the modern world economy by a millennium, and, on the other, to the modern so-called bureaucratic-authoritarian state. World-system theory, no less than modernization theory, offers only a partial, incomplete understanding of Brazil's dependent development, while a secular theory of patrimonialism transcends the anomalies of both theories.

To be quite clear: within the Weberian perspective, the question is not whether Brazil is to be pigeonholed as patrimonial. There is no room for an "essentialist" or idealist account of patrimonial persistence. Nor do I mean to suggest that Brazil has been a static, somehow ideal-typical patrimonial society. To the contrary, clearly there are other than patrimonial dynamics at work. Yet in central ways Brazil has been permeated with patrimonial forms of organization. Thus, the task is to understand how people of power have exploited patrimonial relationships, how they have tried to contend with challenges and external circumstances, the ways in which patrimonial forms have persisted or changed, accommodated to other social patterns or diminished, and what effects these dynamics have had upon Brazil's developmental history. In all this, it will be important to remember that patrimonial power is not one-sidedly hierarchical: all kinds of patrimonial and bureaucratic officials, as well as the independently powerful, have at their disposal ways of undermining rulers and their efforts to maintain control. It is equally important to recognize that Weber's typology of patrimonialism is not a static model; it suggests the interplay of diverse interests under a range of shifting conditions.

Patrimonialism, for Weber, has origins in the patriarchal authority of a family head. When extended to the subordination of other households, patriarchy evolves into a patrimonial relationship of dependency and customary deference on the part of the subordinated individuals and groups, and customary obligation on the part of a patrimonial authority. Yet no patrimonial figure can guarantee allegiance of dependents, who may seek independent power, and so the patrimonialist may seek to ensure allegiance through the distribution of benefits, perhaps coupled to specific obligations on the part of beneficiaries. Thus, there may exist a complex web of such relationships, sometimes overlapping and even conflicting. Any given relationship depends on the capacity of the patrimonial authority to forestall autonomous power and to claim an effective monopoly on resources sought after as benefits. When such monopolistic claims on power and resources can be extended, on whatever basis, beyond the hierarchy of households, and into territory and its inhabitants on a general basis, Weber would speak of the *patrimonial state* as a form of political organization with a distinctive orientation. As he put it, "The patrimonial state offers the whole realm of the ruler's discretion as a hunting ground for accumulating wealth."[17]

Patrimonial states are inherently unstable, partly because external threats can eat up resources in the maintenance of armies, partly because internal organization is based on precarious allegiances that can deteriorate and on authority in principle that can lose its effective claims in practice. When patrimonial claims of authority are totally lacking in force and the ruler depends on extending fiefs in exchange for allegiance, Weber speaks of occidental feudalism, "a marginal case of patrimonialism that tends toward stereotyped and fixed relationships between lord and vassal."[18] Theoretically, there are fluid transitions between various subtypes of patrimonialism, and between patrimonialism and feudalism. However, the typological alternatives do not involve radical differences in the logic of domination; instead, they are based on alternative strategies of maintaining power and appropriating wealth, and on variations in the relative power of a superordinate power and local powers.

Despite the instability of any given patrimonial regime, patrimonialism as a pattern of organization can be quite durable, for it consolidates power in a way that tends to limit the options of opponents, forcing them to seek power within much the same framework. Thus, the collapse of any particular regime likely leaves fragmented patrimonial power networks, such that there may be either a transition toward more "feudal" conditions, or, alternatively, the reestablishment of a state through the emergence of a new ruling patrimonial network. The tendency of competing groups to try to monopolize resources thus channels effective power within the patrimonial framework. Typically, patrimonial rulers better serve their own interests by controlling limited resources than by accepting "progress" that expands societal resources, but loosens their control in the process. In this respect, Weber's analysis of Asian and Arab patrimonial regimes converges with Marx's account of the so-called Asiatic mode of production: both argue the capacity and interest of power holders to suppress or co-opt technical innovations, ideas, activities, and social groups that threaten the established order. Conditions in Europe were decisively different from Asia and the Arab world, tied as they were to an unraveling of the patrimonial empire of Rome toward occidental feudalism that in the long run left an opening for the development of an urban bourgeoisie.[19] But Portugal, in its trading empire and its colonization of Brazil, was not typical of the West.

Patrimonialism in Colonial Brazil

Patrimonialism in Brazil has a lineage dating to the ancient Mediterranean. Both the economic and political frameworks of Brazilian settlement clearly predate the emergence of the capitalist world economy described by Wallerstein, and, once these frameworks were established, they permitted power based on political monopolization of resources to persist through the end of the colonial era in 1822, and up to the present day.[20]

When the Portuguese began settling Brazil in the early sixteenth century, they quickly realized that they could not depend on production by natives, as they had elsewhere in their trading empire. A labor shortage in Portugal, the absence of clear-cut economic gain, and the overextended finances of the crown further meant that the colonization could not pose a drain on state resources. The crown's attempted solution was a form of colonization that maintained the king's absolute right of jurisdiction, but extended a *senhorio*, a grant to a *donatário* of the lifetime right to settle and defend a territorial captaincy. The *donatário* legally could enslave natives, grant licenses for development of collective production facilities such as sugar mills, and use *sesmarias* to deed or rent land to typically wealthy settlers, with the condition that they improve the land. This they tried to do, sometimes ending in failure, but, in certain captaincies, establishing workable colonies based upon use of slaves in the production of sugar for export. Thus began a settlement that in time was to expand its colonial frontiers to the dry uplands of the *sertão*, to the south, and to the Amazon.

How was this colonial history shaped by a patrimonial dynamic? To address that question, I will offer a sociological survey of historiographic research on a series of linked issues: the nature of the colonial political regime, the organization of commodity agriculture, patterns of elite family kinship, elite family relations with the ruling regime, and the character of the developing mercantile capitalist economy.

Patrimony and Administration in Colonial Brazil

We can read the history of Brazilian colonial politics as a series of patrimonial struggles between the crown in Portugal and its subjects in Brazil over the monopolization of resources and opportunities. In this reading, the occasional transformations of politics reconsolidated patrimonial power in new arrangements, rather than bringing any fundamental change from patrimonial to other forms.

The initial colonization of Brazil based on donatory captaincies and *sesmaria* land grants left it to settlers to conquer territory and develop resources and production, while proscribing subinfeudation of power. Ultimate power of granting and withdrawing *senhorios*, ultimate legal jurisdiction, all ungranted lands, and rights of taxation were reserved to the king. Many have argued, Andre Gunder Frank most strongly, that such arrangements cannot be called feudalistic, in the European mold.[21] Yet clearly, too, it is wrong to speak of capitalism in any modern sense. The debate survives as a sterile and simplistic dichotomization between static constructs of "feudalism" and "capitalism" that fail to consider the significance of patrimonialism. If, with Perry Anderson, we recognize that most European feudal regions were structured by a "parcelization of sovereignty," Portugal itself tilted in the opposite direction—toward a patrimonial state in which the more "feudal" parcelization of sovereignty alternated with a strong crown, in a cycle of rivalries among vassal recipients of royal patrimony.[22] In both Portugal and Brazil, we may speak technically of seigneurial proprietorship, in which the right to control land and wealth was allocated politically, while the method of expropriation and use of surplus production were variable, and included production for markets. The latter possibility may be described as "patrimonial capitalism."[23]

To be specific, the initial Portuguese settlement based on the *senhorio* as an instrument approximates what Weber termed "prebendal feudalism": in such a situation the distribution of income opportunities depends on a patrimonial assertion of monopoly control, and is not based on the personal fealty of occidental feudalism, but on fiscal calculations.

In newly claimed territories, the patrimonial ruler simply grants rights to administer and develop land to someone who has to turn theoretical rights into effective power on the ground, while the ruler demands tribute and claims ultimate jurisdiction. In the case of Brazil, this prebendal feudalism involved, as Weber describes it more generally, "the transfer of the risk involved in fluctuating income to an entrepreneur; that is, a sort of tax farming."[24]

Prebendal feudalism, of course, was hardly a product of the capitalist world economy in the sixteenth century. To the contrary, the form of the Brazilian *senhorio* can be traced directly to earlier Portuguese settlement of the Madeira Islands, and to Italian colonization concessions in the eastern Mediterranean in the eleventh century. Indeed, prebendal feudalism was to be found across the ancient world, from China to India and the Mediterranean. In this sense, the colonization of Brazil reasserted a more ancient form of centralized patrimonial seigneurialism against the parcelized feudalism that had dominated medieval Europe.[25]

However, most donatory captains in Brazil did not meet the crown's goals: establishing settlements that would claim territory without crown colonization and generate revenues without crown investment. In these circumstances, the first major political transformation of Brazilian colonization represented nothing more than the reassertion of royal patrimonial prerogatives by the gradual termination of the prebendal arrangements. The less than satisfactory performance of the captaincies, the crown's interest in securing the territories of Brazil against foreign rivals (most notably, the French), and the possibility of monopolizing the increasingly profitable sugar trade led Dom João III in 1548 to establish direct but limited royal control over Brazil through a resident governor-general who would reside in Salvador, in the captaincy of Bahia, established as a capital city. The king could not immediately void all *senhorios* through which rights had been granted to donatory captains, but he moved in that direction, at the same time directing the royal administration to engage more directly in the sugar trade. To do so, the king granted *sesmarias* that contractually established obligations of agricultural development, and he offered tax incentives for building *engenhos* to process sugarcane. The crown taxed sugar and deployed inspectors to monitor its quality with an eye to the European market, it sent slaves to Brazil to be sold as laborers, and it established one *engenho* for the direct profit of the royal coffers. In these ways Brazil came to be more tightly linked to the Portuguese crown's trading empire. Before Brazil, the king had been a patrimonial capitalist ruler who financed his throne through trade carried out between his personally owned fortified trading posts (or factories), which he protected by his navy. In Brazil the crown organized a major colony engaged in direct production as well.[26]

The Patrimonial Dynamic in Colonial Brazil

The reassertion of royal prerogative thus marked a transition from a prebendal feudalism to prebendal patrimonialism. Yet, as James Lang has pointed out, the Portuguese kings established royal authority in principle, but they never completely extended it over the captaincies. Colonial government was problematic for any European power, and perhaps especially so for the Portuguese in Brazil, where royal administration improvised upon the legacy of the overseas trading empire: establish the minimal presence sufficient to tax trade and distribute patrimonial opportunities and grants of land. So long as Brazil engaged primarily in production of agricultural commodities for export, effective royal control over wealth generally could be asserted in the Lisbon harbor and by a presence in the major Brazilian ports, and it did not go much further. Even the discovery of gold and diamonds in the eighteenth century and the development of the interior did not alter the basic strategy; the crown simply extended its authority and its patronage along the same limited lines into new territory. Most telling, even the critical matter of tax collection did not warrant much of a state bureaucracy; instead, the king's exchequer entered tax farming contracts at fixed rates per annum for specified districts; the arrangement offered the crown more prebends to distribute, without encumbering state finances and organization in the collection of revenues.[27]

No doubt the crown failed to maximize revenue on the basis of its approach, especially with respect to gold. This suggests not that the crown failed to act rationally, but that its objectives were patrimonial ones—maintaining its prebendal alliances and its control over the conduits of trade. This approach resulted in a lower level of public revenue than would be realized in theory by opening the channels of trade and establishing a rational bureaucratic administration of revenue collection, but the practical costs of attempting such royal absolutism could not be calculated in advance.

For the Portuguese crown, the enduring problem in Brazil turned on establishing a colonial administration in which the king's authority was not subordinated to local interests—the merchants and the landed agricultural elite. To this end, the crown sometimes employed the usual patrimonial solutions, awarding offices to loyal allies, or selling them for the revenue.[28] It also moved to balance the patrimonial staff lines of control that ran through the captaincies' governors by establishing the crown's own political elite—what might be described as a "mandarin" judiciary, with judges for the *Relação*, the high court, educated at Coimbra in Portugal, limited to fixed terms of office, and theoretically forbidden to conduct business or enter into familial relationships with Brazilians.[29] This rationalistic and nonpatrimonial structure for the administration of justice itself paradoxically would help maintain the crown's patrimonial power, by countering local powers. In any event, the judiciary only slowly penetrated the countryside. Even among the plantations in the Recôncavo of Bahia, the crown's professional magistrates did not appear at the local level until the seventeenth century.[30]

Each with independent ties to the crown in Portugal, Brazil's governor-general, the captaincies' governors, the judiciary, and the treasury were to operate with some degree of autonomous authority that included the monitoring of one another's fulfillment of crown objectives. Beyond this minimalist apparatus of a patrimonial state, the crown could distribute official patronage positions as another basis for nurturing authority. Yet none of these approaches necessarily thwarted contending Brazilian powers. To the contrary, in the major captaincies other than Bahia—Pernambuco and Rio de Janeiro, and even more so in the vast interior—the crown had to recognize that local elites were in control. Contending with conditions of trade in a world economy, the crown operated in a different context than feudalism, yet without effecting broader, absolutist administration of territory.[31] It is the hallmark of a patrimonial state concerned mainly with revenue opportunities that the crown could endure the countervailing power of local elites, so long as it controlled trade, and so long as captaincy governors managed affairs without provoking clear-cut subinfeudation or rebellion.

In short, the crown did not establish a territorial government of absolute sovereignty and civil administration; instead, it strengthened the captaincies as prebends subjected to royal authority, and endeavored to protect royal interests by establishing countervailing patrimonial staffs in the governorship and treasury, and by creating an administrative class of magistrates that lacked a basis of power independent of the king.[32] Given that the prebendal donatory captains and the judicial officials themselves could extend discretionary powers on behalf of the crown, but beyond its effective control, the royal efforts at patrimonial hegemony were structurally undercut in the first instance by the very apparatus designed to maintain them.[33] The colonial apparatus simply established the structure through which the king could attempt to assert his patrimonial claims. These claims were countered by vassals and officials who gained livings by the same patrimonial strategy as the king, but on a narrower scale. They were undermined even further by the classes that organized the production of wealth in the first place, the sugar planters and mill owners (*senhores de engenho*), cattlemen, tobacco farmers, and, eventually, coffee planters.

Agricultural Production as Patrimonial Capitalism
In the earliest important agricultural sector, sugarcane production, the

entire range of relationships—between mill owners and other landowners, landowners and workers, the agricultural elite and government officials—have been the object of enduring arguments about their "feudal" character. Of course they are not feudal in the European sense of a "natural" manorial economy. The question to raise is whether they were structured by a patrimonial dynamic. Against Marx's argument that the economic "base" of a social formation shapes "superstructure," Max Weber held that the economic structures accompanying patrimonialism are indeterminate: "Patrimonialism is compatible with household and market economy, petty-bourgeois and manorial agriculture, absence and presence of capitalist economy."[34] Indeed, patrimonial organization to some extent conditions economic action, because it shapes the nature of economic opportunities.

In colonial Brazil, patrimonialism shaped the whole thrust of agricultural development by a political monopolization of resources that undergirded economic tendencies toward monopolization—most significant, land, the rights to market commodities, the technical means of production, and an adequate supply of labor. In order to survive, landowners faced important considerations other than the economic calculations of production for the marketplace; they were best served financially if they established their own bases of patrimonial power. They were in a strong position to do so because the crown depended on their products for tax revenues.[35] Taken together, the counterposed establishments of patrimonial power—in the state and among the agricultural elite—impeded the emergence of modern capitalism based on free markets in labor, commodities, and land.

The plantation economy of sugar, tobacco, and eventually coffee production in Brazil continued to be based on slave labor long past the end of the colonial era, up until the abolition of slavery in 1888. But even in the case of sugar, where slavery was a fundamental basis of organization, conditions varied considerably, to the point of some slaves obtaining a semi-serf status, paying in kind or in labor for rights of tenancy. Still, patrimonial relations of dependency, not free labor conditions, continued to obtain. Whatever the particular arrangements, they seem best explained as deriving from planter and mill interests in maintaining a coerced labor force, using whatever strategies seemed workable under shifting conditions.[36]

From the colonial period onward, plantation owners' interest in maintaining a dependent labor force, whether slave or peasant, created sometimes paradoxical patron-client relationships. Slaves themselves were part of the patrimony. There are occasional examples of patrimonial motifs in the relations between owners and their slaves, for example, the occasions when an owner would serve as the godfather for a slave child.[37] But even when former slaves obtained the nominally free status of peasants, they tended to remain dependents, insofar as large landowners successfully monopolized land.

Whatever the specific internal patronage relationships of master and slave, landlord and peasant, such relationships were not so crucial to the agrarian structure as the patrimonial principle of resource monopolization that generated them. It was, after all, the crown that held ultimate claim to all Brazil's land; in turn, the *sesmaria* system of land grants created an initial concentrated land distribution that tended to be perpetuated over time, both in plantation areas, and in regions where ranchers established what for all practical purposes amounted to private armies. The Portuguese who received grants of land were not always nobles; prominent in their numbers were military men, professionals, and merchants, including "New Christians"—former Jews whose conversion was forced and therefore suspect. Whatever their social origins, actual settlers of land were not the dispossessed who later fled Europe to the New World. Rather, they were men of some means seeking opportunity, and they aspired to the seigneurial life, as did their Brazilian descendants.[38] Early on, they amassed land in ways that both necessitated and forced laborers, slave or peasant, to depend upon them for the means of production.

In the case of sugarcane, use of land was ordered through complex and conflicting agreements, obligations, and claims, both in sharecropping and rental agreements between owners and producers, and in obligatory contracts for so-called captive cane, in which a *lavrador* (cane grower) was legally required to supply cane to a particular *engenho*. These arrangements amounted to patrimonial control of access to a scarce resource. A key question concerns the consequences for Brazil's position in the world economy. No doubt there are diverse reasons why Brazilian sugar lost out to the West Indies in the world market in the latter half of the seventeenth century. Brazilian plantations could not easily adjust to increases in the cost of slaves and the periodic drops in sugar prices that were caused in part by expansion of production in the West Indies, itself fueled by exclusion of Brazilian sugar from expanding northwestern European markets. Yet it should be noted that Caribbean production increased in the face of falling prices, even as Brazilian production declined. The Caribbean response to such conditions could occur only because of increased economies of scale, and these in part resulted from the consolidation of landholdings into more efficient units of production. In short, the Caribbean sugar economy responded to market conditions in ways that Brazil could not, because patrimonial arrangements of land control held sway in the Portuguese colony.[39]

Just as land monopolization, slavery, and the establishment of a virtually landless peasantry created a coerced supply of labor and forestalled the emergence of markets in land and free labor, in similar ways, monopolization of the technical means of production and of commodity markets created dependency in other arenas of the agricultural export economy. Here again, patrimonialist monopolization of economic opportunities structured arrangements. In the early phases of colonial settlement, the right to build an *engenho*, and thus the technical means of producing sugar, were crown concessions granted by the *donatario*, later by the governor, generally to the richest colonists in an area. The traditional sugar aristocracy did not own all the mills, but its mills were the largest and best situated, and they dominated production: even by 1818, of the 316 *engenhos* in Bahia, 92 were owned by 20 connected families. No doubt their position was based in part on capitalist tendencies toward monopolization, but it also derived from initial comparative advantage that was patrimonially shaped, and maintained through inheritance. Major production, and with it, control of the industry, tended to concentrate in the hands of a small number of great families.[40]

Whatever their social status, mill owners sought to use their positions in the production system, their capacities to extend credit, and their positions as landlords to obligate "captive" tenant and sharecropper producers of cane to ensure a steady supply to their mills. Even beyond their own lands, the *senhores de engenho* acted to create monopolies by exploiting the relative absence of alternative sources of credit to growers. Similarly, the crown could favor one or another colonist with credit on the basis of considerations that did not necessarily have anything to do with the economics of cane production. Brazilians—in this case, planters—thus could use the state for their own ends, even if they depended on patrimonial discretion to do so.

The capacity of the crown to intervene on a noneconomic basis injected an incalculable element into commodity production. Even as such intervention benefits one group or another, the sort of incalculability at work strengthens a patrimonial regime's hand, for uncertainty forces political subjects to cultivate relationships with the regime in hopes of favorable treatment. By the same token, incalculability inhibits increasing predictability of operational conditions, which Weber insists is a cornerstone of capitalist development. Thus, what might be dismissed as inconsistent policy from a capitalist point of view may make sense under a different rationale—the interest of the crown in maintaining circumstances in which others have to seek its favor.[41]

Practices of political favoritism, credit patronage, and "captive" cane agreements undermined the emergence of free markets in either commodities or credit, and cane growers fought an ongoing (and apparently losing) battle to approach even the oligopsonistic conditions in which they would have had more than one buyer from which to choose. To be sure, both growers and mill owners operated for profit, but they did so in a world where noneconomic controls and politically arbitrary considerations could structure conditions of economic action.

To further understand patrimonial agriculture in Brazil, it will be important in future research to compare other commodities, over time, with sugar. Whatever the findings of such research, it seems evident that Brazilian colonial plantations and ranches existed in a world where capitalist interest converged with an interest in protection against capricious exercise of state power. Capital was useful for cushioning an enterprise against the short-term ups and downs of markets, but, as elsewhere, pursuit of property, wealth, and power were the ways to endure the long waves of downturns in world trade. In Brazil, these considerations also weighed heavily in the capacity of the landowning class to counter the potentially arbitrary patrimonial state.

Patrimonial Wealth as the Production of Kinship in Elite Families

Beyond the ways that patrimonial relations structured agricultural commodity production, the significant issue about the agricultural elite concerns the amalgamation of wealth itself, and its transfer from generation to generation. Here, patrimonial power flowed from the capacity to monopolize scarce and sought-after family resources.

It is my contention that elite agricultural families did not typically seek out other opportunities for investment to maximize profit in capitalistic terms. They were more keenly interested in amassing wealth to maintain a seigneurial style of life and position in society, from generation to generation. Landholdings, not profits, were the salient measures of status: "The successful merchant proved his worth by abandoning trade." The agricultural elite, not an independent merchant and business class, was socially most prominent and powerful; only in Pernambuco did merchants maintain a viable elite independent of the more powerful planter elite.[42]

Though it has been difficult to conduct research on genealogies and family wealth for the colonial period, an interesting picture emerges from what has been done, and from roughly comparable analyses for later periods. Essentially, among traditional elite families, the politics of marriage and inheritance tended to be dictated by considerations of maintaining and concentrating family wealth, by way of a number of pragmatic strategies. As in Europe, "excess" daughters could be sent off to convents, as a way of avoiding the fragmentation of family wealth in expensive marriage dowries.[43] Even kinship itself could be reckoned on grounds that concentrated wealth. Thus, the family as a patrimonial unit did not include all kindred relatives of an extended family, and descent could be traced through either spouse of a given descent couple. This "volitional" and flexible approach made it possible to bypass kindred family lines that might dilute the concentration of wealth or threaten the control of it. In a parallel way, the tendency toward endogamous marriages, of first cousins or uncles and nieces, solidified alliances within an "operational" family that was narrower than the extended kinship descent group. Such strategies not only concentrated wealth; they gave rise to conflict within extended families over control of patrimonial wealth.[44]

Exogamous practices operated to the same end, concentrating and controlling wealth, but by the opposite strategy. For example, large landholding families in two regions of the same captaincy might intermarry over several generations to establish an alliance, as the Araújos and the Feitosas did in Ceará.[45] In other cases, elite colonial Parnaíba families used the practice of bringing a son-in-law into the family to establish connections with the merchant wealth and political power of the citics, both in Brazil and Portugal. While such matches offered social mobility for merchants and other individuals, effectively, they reinforced the position of the agricultural elite family as a key institution of Brazilian society.

The basic thrust of family strategies was strong enough to lead Alida Metcalf to conclude, "it would appear that the origins of the Parnaíba class society are more to be found in the strategies of elite families than in external economic factors, such as the expansion of commercial agriculture."[46] If this is an overstatement, it still forces us to recognize a different process at work than the holistic determinism of a world economy: however much elite families had to contend with economic conditions, they operated in pursuit of goals autonomous from straightforward pursuit of profit through production or trade. Here the issue is not the existence of a titled nobility that constituted a bounded estate: as Schwartz has shown, the social origins of the wealthy were diverse, and there was some degree of upward and downward mobility.[47] What matters is that families operated as status groups that sought to monopolize resources. Meaningful kinship was a basis of social position that competed with class position. Outside the framework of economic production, families sought to become and remain powerful. They did so not only within the framework of kinship, but in the political sphere as well.

Families, Patronage, and State Patrimonialism

In Max Weber's view, the ruler of a patrimonial state faces dual pressures in the development of administration. On the one hand, he seeks to maintain power independent of his subjects, for it is on this basis that he avoids a slide into feudalistic conditions. As we already have seen, this interest may even lead toward establishment of a rationalistic bureaucratic organization structured to counter the potential subinfeudation based on prebends. On the other hand, the ruler tries to avoid a complete extension of civil authority and the cost it entails. Moreover, the careful distribution of benefits can help to maintain loyalty. Thus, efforts toward bureaucratic rationalization of officialdom are balanced by a need to use patrimony to accommodate local power interests.[48] For their parts, colonial elites depend on predictable trade. Therefore they support in principle an independent and legitimated source of authority. Still, it is to their advantage to influence and control that authority wherever possible.

In Brazil, these considerations, not to mention the financial benefits, mark the efforts of colonial elites to control offices, patrimonial or bureaucratic, and to influence officials by cultivation of mutual interests. Whatever the goals of local power groups, they framed such goals within a patrimonial logic. Thus, we have the odd spectacle of colonists who both attacked corruption, presumably when it did not go their way, and in turn might defend the most corrupt officials.[49]

Stances toward crown power on the part of the Brazilian landholding elite varied, both within regions and between them.[50] It is also the case that the merchant classes in some captaincies, such as Minas Gerais, operated with greater autonomy than in those captaincies centered on agricultural exports.[51] Whether the landowners' alliance with the merchant class was strong or weak informed the contents of political agendas, but in either case the vehicle was largely the same: the pursuit of position and influence within the patrimonial and bureaucratic apparatus of the crown.

In part the crown attempted the establishment of "mandarin" authority independent of local powers through the *Relação*, the high court.[52] Perhaps the crown sought to reinforce dependence of this elite on Portugal by choosing among Brazilians mainly from middle echelons of society, especially a self-perpetuating bureaucratic class, rather than from among the sons of the most powerful. Yet autonomy was undermined in a number of ways. The efforts to maintain a quarantined class meant less, once Brazilians attended Coimbra in numbers, for there they formed the personal elite friendships that helped consolidate the networks of local power on the home front when they returned to serve on the high court.

Nor did other crown efforts to preclude corruption in Brazil prove entirely successful, for justices were both human and powerful. Landowners seldom directly sought judicial positions for themselves; presumably they had fatter fish to fry, and most of the Brazilian-born men of the colonial high court came from the middle strata, seeking prestige and upward mobility. But both immigrant and Brazilian-born judges might seek Brazilian wives, and by this forging of kinship they often became connected to powerful families. Schwartz has argued that these are simply the most visible historical traces of linkages between authorities of the crown and the local elite. Where such relationships existed, solidification of family connections, business dealings and partnerships, subordination of state administration, and out-and-out bribery and graft likely became part of the equation.[53]

This is not to say that the Portuguese crown failed in its colonial administration: to the contrary, it created an apparatus that countered local power, in part by the bureaucratic logic that sought to recruit professionals rather than favorites at higher echelons of administration. The bureaucratic officials sometimes effectively pushed toward systematization of policy, and even if they never came close to eliminating the traditionally patrimonial side of Portuguese rule, the bureaucratic tendencies probably increased over time. But the crown's interest in extending this bureaucracy ultimately was to protect its own patrimony, and the distribution of bureaucratic offices itself could be based on patrimonial considerations.

The high court was the crown's best hope of an independent source of authority, yet it also was an object of elite family influence. As we would expect, such influence was far more pronounced both at lower judicial levels and in more patrimonial domains of the colonial administrative apparatus and local government, where jobs as notaries, clerks, and inspectors could be distributed. Commoners may have sought positions purely for the income, and the crown may have used patronage to cultivate their allegiance. If the crown followed the logic described by Weber, it appointed commoners to offices because they had no independent basis of power, and thus could be more easily controlled.

Landowners were another story: because they projected power independent of the crown, they could seek to control offices, and in turn parlay them to further advantage. The control of colonial government by powerful families may be presumed to have been most extensive in the individual captaincies and at the local level. Much of the settlement history of Brazil, especially outside Bahia, could be written as the struggle between competing families and the crown over appointment of governors and circuit judges (*ouvidores*).[54] At the local level, it was simply a competition among elite families for control over all posts, military, bureaucratic, judicial, or representative. As Stuart Schwartz put it, "The utilization of office in the municipal councils and local judicial hierarchy enabled these kindreds to exercise control over many areas of policy and economic activity."[55]

The Brazilian colonial elite, especially the large landowners, but also the more established merchants, engaged in the patrimonial appropriation of local, regional, and colonial positions of power directly or through marriage. In part, to be sure, there were sometimes personal and purely financial motives—the gaining of a livelihood, protecting a destitute *fidalgo*, obtaining tax farming rights. But intermarriage through daughters, the placing of younger sons without inheritances, the holding of multiple offices concurrently, and the absence of bureaucratic careers of promotion—these aspects also encompass an additional rationale: the colonial elite sought to "capture" power resources within the kinship frame of their families.[56] Just as elite families operated internally to maintain patrimony, they sought to establish themselves in the public domain as power organizations in their own right.

Class analysts will point out that local patrimonial societies infused with "patron-client" relations nevertheless mark a particular situation of class power, in which one class is capable of mediating the distribution of resources.[57] The basic point is correct, but that is not the whole story. It is particularly under conditions of patrimonial regimes that local elites are capable of monopolizing resources, and within local landholding families, the competition for class eminence follows a logic of status group competition among families. Thus, patron-client relations are mapped onto social classes, but on the basis of a patrimonial dynamic that cannot be derived from class circumstances alone.

Sociologists like to ask what ruling classes *do* when they rule. In the case of Brazil, the colonial elite were not always most concerned with protecting interests that derived from their own class situation. Co-optation of judicial and governing power often served a simple, yet fundamental goal of protecting the family itself from prosecution. The most powerful family empire within a territory, perceived as not fully subject to the external law, effectively could operate as a "state within a state"—a power that only rarely could be challenged by competing elite families, much less dependents, commoners, slaves, and outsiders.[58] The corruption and graft that dot the history of colonial Brazil's elite family history are simply corollaries of patrimonial power.[59] Yet their prevalence suggests that patrimonial power itself in some cases shaped "class" opportunities of officeholding and public power benefits that did not derive from the mode of production itself.

In countering Portuguese power, the colonial elite families established their own realms of patrimony. Such an accommodation is hardly unusual, in a comparative sense. Rather, it is the bargain that patrimonial rulers accede to, as the cost of maintaining their prerogatives on a broader level. These royal prerogatives were most concerned with taxing

The Patrimonial Dynamic in Colonial Brazil 77

trade, and here again, there was a tension of conflicting patrimonial interests.

Mercantilism and Capitalist Development in a Patrimonial Regime

When trade occurs under patrimonial power, the critical questions of long-term development have to do with the conditions under which freer trade may emerge. As Weber remarked, while feudalism is hostile to trade, patrimonial regimes may secure limited and monopolized trade as a basis for financing the throne. Such arrangements proceed on the basis of decree and concession that lack the calculability needed for modern capitalist enterprise. But, Weber adds, under sufficient impetus to raise revenues and organize monopolies rationally, a transition may take place from highly regulated colonial licensing to relatively freer exchange.[60]

From the beginning, the Portuguese crown monopolized and controlled the Brazilian commodity export trade, first, by establishing crown factories for trade; second, by selling licenses and contracts for merchant activity and tax farming on the slave trade; and third, by attempting to limit foreign powers from trading directly to Brazil. As early as the mid-sixteenth century, however, Portugal was becoming little more than an entrepôt for the trade networks being consolidated from northwest Europe. By the end of the century, Portuguese sugar merchants operated as "front men" in the Dutch move to capture the Brazilian sugar trade, an effort that led to the Dutch armed occupation of Recife from 1629 until 1654.[61]

Crown strategies once again show a consistent pattern only if we understand the patrimonial consistency of inconsistency. The hand of crown favor often tipped to the commodity producers themselves, as is suggested by the tax incentives for mill renovation and the passage of laws that, at least for a time, protected growers from foreclosure to collect on debts. From the crown perspective, merchant-creditors would have to exercise patience: an indebted producer was just as good a source of tax revenue as a solvent one, and far preferable to a failed one. So far as trade itself was concerned, the crown favored the larger growers in 1629 with a decree that required shippers to reserve a third of cargo space for direct consignment sales by producers. Merchants had to be licensed by the crown, and ships bound for northern Europe were expected to pass through Lisbon to pay duties.

In both Portugal and Brazil, the crown would want to prevent the merchant class from obtaining independent power. But the king also depended on them. The solution was co-optation. It was the Dutch occupation that led Portugal to a significant modification of trade arrangements. With Brazil's trade threatened by piracy, the crown founded the Brazil Company in 1649 with capital from merchants, New Christians among them. Drawing on private investors, the company was the first significant move away from a directly administered patrimonial regime. Like competing Dutch and English companies that had similar patrimonial charters, it initially operated with a degree of autonomy from the crown, even though the rights of the company essentially amounted to patrimonial benefices. Investors profited from opportunities dispensed by the crown—tax-farming positions and liens on state revenues—and the company received a monopoly on imports of wine, flour, and other foodstuffs to the colony, with the right to set prices for them. For protection against the Dutch, in the 1660s Portugal sought the help of England, in exchange for trade concessions that further cut into the crown's holdings and confirmed Portugal's position as an entrepôt for trade centered elsewhere. Even if these steps did not always meet the needs of producers who depended on reliable access to European markets, even if Brazilians objected to the price-fixing on imports, from the patrimonial viewpoint of the Portuguese throne, the company succeeded: it monopolized shipping against foreigners and no doubt eased the task of collecting duties. Success nourished the interest of the crown: by 1694, the company was absorbed into the Portuguese government as an agency, and later it became a part of the treasury.

The initial steps toward private investment and subsequent bureaucratization of the company as a state agency moved somewhat toward rationalizing the Brazil trade, albeit within the framework of state domination. But the Portuguese crown still operated according to a patrimonial logic that was antagonistic to independent capital.[62] Though the state embarked on capitalist production schemes in Portugal, these halfhearted state capitalist efforts at import substitution were quickly abandoned at the beginning of the eighteenth century, when the revenues from trade picked up. In Brazil, the Portuguese crown retained control of shipping and excluded English merchant factories, even though the English took over the Dutch role in financing traders and supplying manufactured goods.[63]

The reemphasis on trade meant that capitalist investment had to follow the king. Thus, it is true that the entrepôt position of Portugal reinforced dependency, but it is important to recognize the way this occurred. The original colonization of Brazil had established revenue opportunities along patrimonial lines *before* Portugal became an entrepôt, and in ways that predated the emergence of the modern capitalist world economy. This situation did not change after Portugal lost its predominant position in the world economy at the beginning of the seventeenth century. Subordinate ties to the mercantilist economy made fixed

The Patrimonial Dynamic in Colonial Brazil

capitalist investment a less attractive proposition for the crown than a strategy of revenue spoils patrimonialism. The logic of patrimonialism was not undercut by trade in the case of Portugal; it was strengthened.

Not until the Marquis de Pombal became the king's minister of state in 1750 was there a serious challenge to the patrimonial organization of trade and the Brazilian economy. Until then, the crown's approach to the colonies was concerned with generating revenues to finance the court, and Portugal's merchant class failed to prosper because it lacked either capital or political clout independent of the crown.[64] Perhaps the continent's first dependency theorist, Pombal wedded a keen sense of international trade to the power of state patrimonial capitalism. By organizing Brazilian monopoly companies along aggressive capitalistic lines, he sought to decrease dependence on English imports. The plan was not to diversify Brazil economically, but to create a source of capital to finance import substitution production in Portugal. In a single stroke, Pombal managed to reinforce the position of Brazil's agricultural elite, revitalize a Portuguese business elite directly tied to the crown's fortunes, and rationalize the colonial fiscal administration while maintaining Brazilian elite and merchant participation in its profit opportunities.[65]

With Pombal's regime, state patrimonial capitalism took a modern turn. His program and subsequent variations on it were not without opponents in Brazil, particularly among those who might benefit from the development of Brazil's internal economy and from free trade that bypassed Portugal.[66] But at the turn of the nineteenth century, a boom in Brazil's export economy made these initiatives moot.

In a much more permanent way, the relocation of the Portuguese court to Rio de Janeiro in 1807–1808 under Napolean's threat shifted the terms of the patrimonial equation. The Portuguese crown no longer would seek to block free trade, for trade generated revenues, and with the crown in Brazil, there was no substantial reason to protect merchant and manufacturing interests in Portugal against Brazilian incursions. Yet these changes did not reinforce the halting steps in Brazil toward import substitution and growth of the internal economy, for the Portuguese king had to turn to Great Britain for protection by the Treaty of 1810, undercutting the Brazilian merchant class by further enhancing an already privileged British competition. In the bargain, the British moved against the Brazilian slave trade, creating a split between planter interests and the political needs of the crown.

Again, as with the earlier opening of Brazilian trade, the ironic result of reintegration into the world economy was to strengthen the position of the agricultural landowning class. The weakness of the Portuguese regime in the world economy was predicated upon its antiquated patrimonial structure, and that weakness in turn reinforced the position of the Brazilian patrimonial landowners and their relatives and elite allies at the dawn of the industrial era.

Ultimately, the Portuguese crown's dominion over Brazil ended through a decisive split in the patrimonial interests of Portugal and Brazil, interests that had so long meshed, even in their opposition to one another. The king returned to Portugal in 1821 to legitimate his claim to the throne, leaving the crown prince Dom Pedro in Brazil. By then Portugal would want to reestablish the patrimonial regime with Brazil as colony, but it lacked the power to enforce that result on its former realm of patronage: Dom Pedro declared the independence of Brazil in 1822, and set about to solidify his position as the head of a state whose basic structure already had been established.[67] The long-term monarchical power interests that Max Weber cites as favoring the preservation of territorial boundaries under patrimonial regimes could not prevail over pressures for division.[68] But this collapse of the king's authority did not have the immediate result of breaking the patrimonial mold. Instead, the independence of Brazil amounted to a hereditary division of patrimonial domains that substituted an emperor for a king who was his father.

The Legacy of Patrimonialism in Postindependence Brazil

After independence, mercantilist modernization of the state and the strengthening of the already more diversified internal economy in Rio de Janeiro created new cleavages in Brazil. These developments transcended the purely patrimonial dynamic of conflict over spoils between a royal court and the elite of its colony. Yet the postindependence changes did not eliminate patrimonial structures. Instead, the institution of a constitutional monarchy and the abdication of Dom Pedro in favor of a regency in 1831 created a new patrimonial problematic in the relationship between the state on the one hand and local powers on the other. Permeated with officials of Portuguese allegiance, the state lacked effective legitimacy. An opening was thereby created by which local powers could tip the state/local balance to their own advantage. Yet national power abhors a vacuum. The liberal era of the regency was not only a time of local and regional movements toward decentralization, it witnessed the reconfiguration of a national elite that came to occupy and modify the administrative structure vacated by the former colonial power.[69] To be sure, elections were increasingly invoked as a basis of representation, and the patrimonial state continued to develop a bureaucratic administration that paralleled its patrimonial networks. But in Weber's analysis of patrimonialism, these events hardly constitute a revolution, or even reform. Instead, they seem to map an increasingly complex society onto relatively enduring patrimonial institutions that maintain a tension between local and regional powerholders and an encompassing administrative regime.

In charting the patrimonial legacy of the colonial era, I have sought to show that a distinction between feudalism and the administrative state offers a false opposition. When a regime seeks to establish and maintain territorial power without the capacity or will to administer local affairs, patrimonial and bureaucratic initiatives offer alternative strategies in the patrimonial ruler's struggle to counter local power. The patrimonial dynamic seems to be maintained so long as neither the ruler nor local powers win a decisive and irreversible victory.

The classic study of Caio Prado Júnior asserted that Brazilian independence marked the exhaustion of the colonial endeavor and the beginning of the "contemporary phase."[70] There is a long-standing debate about whether in this new phase Brazil did or did not break from its past. I have thrown into sharper relief the patrimonial dynamic of that past. It is not possible here to offer even a schematic typological analysis of postcolonial Brazil in the same terms.[71]

For the moment, we will have to make do with sketching hypotheses about the patrimonial dynamics that have shaped "dependent development" of Brazil.[72] First, from the empire to the present, the dynamic of patrimonialism would suggest a continuing battle to impose national rule over and against the discretionary power and patronage orientation of local notables, who themselves sometimes sought national power. This issue forces a second: how successfully have patrimonial capitalist landowners forestalled the emergence of free markets in capital, land, and labor that would mark a transition to other structures of capitalist agriculture? On the other hand, what are the conditions that have permitted capitalist industrial classes to emerge, especially in the center-south, and how have these changes shifted the institutional basis of patrimonial power? How have other nation-states and non-Brazilian companies accommodated with patrimonial realities, and to what extent have they undermined them?

Within the state itself, there are a number of developmental issues. We would want to know where universalistic bureaucratic administration effectively has challenged the patrimonial ethic of patronage. Has state patrimonial capitalism of the ancien régime offered an institutional basis that fueled the recent bureaucratic-authoritarian regime? To what extent has the patrimonial logic of revenue appropriation and patronage pushed development along state capitalist lines, as opposed to alternative solutions, for example, greater accommodation with independent capitalist corporations? Finally, how did the patrimonial dynamic of colonial Brazil structure the particular path of nation-building through limited-franchise electoral politics and corporatism?

Much already has been done to explore the patrimonial features of Brazil since independence.[73] It is widely recognized, even by those who advance "externalist" explanations of dependency, that entrenched social classes and state officials have successfully maintained their "backward" pursuit of patrimony precisely on the basis of their capacity to accommodate and even redirect capitalist initiatives.[74] Yet an overall analysis of postindependence Brazil has not yet been fully consolidated. The present survey suggests that patrimonialism in the colonial period had developmental dynamics that cannot be reduced in any straightforward way to the causal dynamics of a world economy or world capitalist accumulation. If the same is true for Brazil from the empire onward, those strata seeking to shift the unequal development of Brazil will find that understanding the patrimonial dynamic suggests different points of leverage than purely economic policies.

For Brazilians who are not the direct beneficiaries of patrimonial arrangements, there is no single solution to the patrimonial domination of life. For it may involve a coherent dynamic, but it is a diffused and extensive form of power, not a purely centralized and intensive one.[75] In this way, patrimonialism amounts to a cultural structure that has permeated the social life of Brazil on multiple fronts.

Conclusion

The world-system approach as a master theory of history may appeal to some historians because it seems to offer a convenient model that gives shape to more focused empirical studies. But the appeal is unfounded for several reasons. In the first place, universal histories rarely are adequate to their own explanatory ends. Second, they unnecessarily narrow the agenda of historiography. Third, they fail to contend with the methodological problems of historical inquiry, substituting a philosophy of history for strategies of empirical analysis. These difficulties are not visited upon every study that claims the world-system perspective as its provenience, for the perspective includes eclectic approaches that employ other methods and depend on other assumptions to address world-system questions. We need to remember that actual research often is either better or worse than the theory that sustains it.[76] Yet practitioners are faced with a basic choice that echoes across all philosophies of history: insofar as the world-system perspective offers theoretical explanations, in the strong sense of that term, empirical assessments may find the theory wanting, and give grounds for displacing the edifice altogether. If, on the other hand, the approach amounts to a sensitizing perspective rather than a comprehensive explanation, it loses the cohesive framework that differentiates it from other historical inquiries; only the distinctive conceptualizations and questions about relationships between key phenomena remain, and they may be subject to answers from foreign quarters.

If the present analysis of patrimonialism shows anything in a general sense, it shows that an assumption of holism within the world-system perspective is inadequate to an explanation of historical change, for the patrimonial dynamics that are critical to historical explanation—in the case at hand, of colonial Brazil, its antecedents, and its legacy—do not derive from the functional logic of the world system and accumulation on a world scale. As Aronowitz has noted, most substantive critiques of world-system theory fail to address the theory's assumptions, and thus miss their marks.[77] The present study has proceeded in a different fashion, to show the relevance of an alternative approach for explaining events, an approach that challenges the holistic assumption. Once holism is abandoned, the warrant of world-system theory as a master theory of history no longer has force, and the world system becomes a level of analysis, with accumulationist theories forced into contention with other explanations, be they neo-Marxist, neo-Weberian, or simply historicist in the best sense of that much abused term. In substantive theoretical terms, it would be naive to think of capitalism as some pervasive holistic system that, once born, functionally defines all participants by their relations to it. Surely the opposite is true, that individuals, groups, and classes orient to capitalism in different ways, and in doing so, shape the nature of capitalism.

Weber's sociohistorical typology of patrimonialism describes an array of possible social relationships. For any empirical set of patrimonially organized relationships, the lines of articulation with other complexes are diverse, as are the possible lines of transition to other structures. Any reductionist attempt to subsume these possibilities in an economistic theory is doomed; any attempt to explain patrimonialism purely in terms of a holistic scheme similarly will fail. For the transitions are often paradoxical. On the other hand, it would be equally wrong to claim patrimonialism as some sort of essence that can be explained independent of the forces that come into play around it. In this survey, I have explored the interaction between patrimonial dynamics and other forces. As I have sought to demonstrate, a developmental history of patrimonial institutions can pull together strands of history that otherwise seem merely juxtaposed, like so many pieces of a puzzle that don't quite fit.

Notes

I am very grateful to Richard Graham, Fernando Novais, George Primov, and the symposium participants at the University of Texas at Austin for sharing their reactions to an earlier version of this essay.

1. Frank, *Capitalism and Underdevelopment in Latin America*. On the reinvigoration of historical sociology, see Skocpol, *Vision and Method in Historical Sociology*. For a review of the emergence of dependency and world-system theory, both before and after Frank's study, see Chirot and Hall, "World-system Theory," 81–106, and for later developments, see Walton, "Small Gains for Big Theories," 192–201.

2. For reviews of the burgeoning literature, see Mörner, Fawaz de Viñuela, and French, "Comparative Approaches to Latin American History," 55–89; Zirker, "Brazilian Development," 135–149; Mauro, "Recent Works on the Political Economy of Brazil in the Portuguese Empire," 87–105; and Graham, "State and Society in Brazil, 1822–1930," 223–236.

3. For a lament on the controversy: Himmelfarb, *The New History and the Old*. An embrace of diversity is offered by Tilly, *Big Structures, Large Processes, Huge Comparisons*. For a deconstructionist critique of historical discourse, see Cohen, *Historical Culture*. On the narrative controversy in particular, see White, "The Question of Narrative," 1–33; Veyne, *Writing History*; Stone, "The Revival of Narrative: Reflections on a New Old History," 3–24; and Hobsbawm, "The Revival of Narrative: Some Comments," 3–8.

4. One solution was the approach of Ranke; see Krieger, *Ranke: The Meaning of History*. On the problem of selection, see Atkinson, *Knowledge and Explanation in History*, 69ff.

5. World system as level: Tilly, *Big Structures*, 63. For relatively eclectic positions, see F. H. Cardoso, "The Consumption of Dependency Theory in the United States," 7–24; Mukherjee, "Commentary on Robert L. Bach," 314–316; Chase-Dunn, "Commentary on Robert L. Bach," 312–313; Mandel, *Long Waves of Capitalist Development*.

6. Wallerstein, *The Capitalist World-Economy*, 53–54, 68, orig. emphasis. See also Bach, "On the Holism of the World-systems Perspective," 289–310.

7. A scathing Marxist critique is offered by Howe and Sica, "Political Economy, Imperialism, and the Problem of World System Theory," 235–286. Perhaps the classic formulation is given by Brenner, "The Origins of Capitalist Development," 25–92. For a detailed orthodox Marxist analysis of accumulation theories, see Weeks and Dore, "International Exchange and the Causes of Backwardness," 62–87; see also: Aronowitz, "A Metatheoretical Critique of Immanuel Wallerstein's *The Modern World System*," 503–520; Luton, "The Satellite/Metropolis Model," 573–581. For other critiques, see Skocpol, "Wallerstein's World Capitalist System," 1075–1090; Halperin-Donghi, "'Dependency Theory' and Latin American Historiography," 115–130; Hall, "World-system Holism and Colonial Brazilian Agriculture," 43–69. Elaborations are diverse; see Booth, "Marxism and Development Sociology,"761–787; a review by Trimberger, "World Systems Analysis: The Problem of Unequal Development," 127–137. Chirot and Hall, "World-system Theory," 97–102, review some of these objections. A response to Wallerstein's critics, especially Brenner and Skocpol, tries at reconciliation by maintaining the "totality" of the world system, while claiming primacy for neither economic nor political variables; see Garst, "Wallerstein and His Critics," 469–495.

8. The potential of dependency theory was mentioned by Graham and Smith, "Introduction," xiii, and noted by Love, "An Approach to Regionalism,"

137–155. Topik, *The Political Economy of the Brazilian State, 1889–1930*, and Schwartz, *Sugar Plantations in the Formation of Brazilian Society*, both use Wallerstein as a foil for their investigations. For explanations of the slow adoption rate, see F. H. Cardoso, "The Consumption of Dependency Theory," 9, 12; Halperin-Donghi, "Dependency Theory," 121–122.

9. The key figures in bypassing Parsons's version of Weber were Reinhard Bendix, who followed Weber's strategy as a comparative historical sociologist, and Guenther Roth, who, along with Claus Wittich, translated and edited the complete version of Weber's magnum opus, *Economy and Society*. For their general approach, see Bendix and Roth, *Scholarship and Partisanship: Essays on Max Weber*. For more recent formulations, see Roth, "History and Sociology in the Work of Max Weber," 306–318; Bendix, *Force, Fate, and Freedom*; Roth, "Rationalization in Max Weber's Developmental History," 75–91. For Schwartz's brief comments on Weber, see his *Sovereignty and Society in Colonial Brazil*, xx–xxi, and "State and Society in Colonial Spanish America," 8–9.

10. Weber, *Economy and Society*, 20.

11. Turner, *For Weber*, 250, takes the "structuralist" view of Weber's concepts, but does not address the ways in which Weber's basic concepts and types are formulated on the basis of an assumption that social action is in principle comprehensible.

12. Roth, "Sociological Typology and Historical Explanation," and Roth, "History and Sociology"; Hall, "Where History and Sociology Meet."

13. See, for example, Buss, *Max Weber in Asian Studies*; Collins, *Weberian Sociological Theory*; Antonio and Glassman, *A Marx-Weber Dialogue*; Lash and Whimster, *Max Weber, Rationality, and Modernity*; Schluchter, *Rationalism, Religion, and Domination: A Weberian Perspective*.

14. Aside from *Economy and Society*, core aspects of Weber's analysis are contained in Weber, *The Agrarian Sociology of Ancient Civilizations*, and Weber, *General Economic History*. On the uses of Weber's approach in the study of the world economy, see Roth, "Personal Rulership, Patrimonialism, and Empire-building," 156–169; Turner, "Weber and the Sociology of Development," chap. 8 of his *For Weber*; Turner, "The Structuralist Critique of Max Weber's Sociology," 1–16.

15. Best known are those to do with the culture of civilizations, most notably, the significance of inner-worldly asceticism for capitalist development in Europe and North America. The emphasis on this supposedly "idealist" strand in Weber's argument sets up a straw-man version of his position in the attack by Frank, "Development and Underdevelopment in the New World," 431–466.

16. Roett, *Brazil: Politics in a Patrimonial Society*; Uricoechea, *The Patrimonial Foundations of the Brazilian Bureaucratic State*; Faoro, *Os donos do poder*.

17. Weber, *Economy and Society*, 1010ff., 1099.

18. Ibid., 1070.

19. For a review of the convergence and a summary of the "stagnation" thesis, see Turner, *For Weber*, chap. 8. For studies addressing similar phenomena, see Eisenstadt, *The Political Systems of Empires*, and Eisenstadt and Lemarchand, *Political Clientelism, Patronage, and Development*. On the Roman Empire and European feudalism, see Anderson, *Passages from Antiquity to Feudalism*.

20. Hall, "World-system Holism and Colonial Brazilian Agriculture"; on the

early phases of Portuguese colonization, see in particular Mauro, *Le Portugal et l'Atlantique*. The narrative account upon which I overlay analysis of patrimonialism draws especially on Lang, *Portuguese Brazil*, both because Lang offers the most recent broadly grounded scholarly history of colonial Brazil, and because he devotes considerable attention to the issues of political and territorial domination upon which a typological study of patrimonialism must depend.

21. Frank, *Capitalism and Underdevelopment*, 143–218. The opposite stance still is argued with passion, for example, by Romano, "American Feudalism," 121–134. The classic analysis is by Marchant, "Feudal and Capitalistic Elements in the Portuguese Settlement of Brazil," 493–512.

22. Anderson, *Passages*, 148; Lang, *Portuguese Brazil*, 4.

23. Weber, *General Economic History*, 51–64; Hall, "World-system Holism and Colonial Brazil."

24. Weber, *Economy and Society*, 1102, 260.

25. Johnson, "The Donatory Captaincy in Perspective," 203–214; Verlinden, *The Beginnings of Modern Colonization*; Weber, *Economy and Society*, 260–261.

26. Lang, *Portuguese Brazil*, 24–29; Schwartz, *Sugar Plantations*, 19–22; Mauro, *Le Portugal et l'Atlantique*, 202, 219ff.

27. Lang, *Portuguese Brazil*, 30, 54–55, 73, 123–131, 147–150.

28. Alden, *Royal Government in Brazil*, 294.

29. The term *mandarin* is used for the empire period by Pang and Seckinger, "The Mandarins of Imperial Brazil," 215–244; it applies equally, if not more, to the colonial period; Weber, *Economy and Society*, 1038–1040.

30. Lang, *Portuguese Brazil*, 42, 46, 67.

31. Schwartz, *Sovereignty and Society*, 215–216; Dutra, "Centralization vs. Donatorial Privilege," 19–60.

32. Lang, *Portuguese Brazil*, 61, 72.

33. Cf. Weber, *Economy and Society*, 258, 1026.

34. Ibid., 1091.

35. On favorable treatment by the crown, see Lang, *Portuguese Brazil*, 85–86.

36. Furtado, *The Economic Growth of Brazil*; Schwartz, *Sugar Plantations*, 193–194; Taylor, *Sugar and the Underdevelopment of Northeastern Brazil, 1500–1970*, 46.

37. Schwartz, *Sugar Plantations*, 64.

38. Lang, *Portuguese Brazil*, 68–70; Schwartz, *Sugar Plantations*, 265–267.

39. Schwartz, *Sugar Plantations*, 183–185. Lang, *Portuguese Brazil*, 110.

40. This and the following paragraph are based on Schwartz, *Sugar Plantations*, 22–26, 97, 199, 209–211, 271.

41. Weber, *Economy and Society*, 1095.

42. Lang, *Portuguese Brazil*, quote, 106, in describing "New Christians"; 146–147.

43. Ibid., 132; Schwartz, *Sugar Plantations*, 290.

44. Lewin, "Some Historical Implications of Kinship Organization," 262–292; Lewin, *Politics and Parentela in Paraíba*; these studies concentrate on the nineteenth and early twentieth centuries, and Lewin notes the ways in which local electoral politics altered the specific strategies that had been important since colonial times. On the tendency in colonial times toward endogamy, see

The Patrimonial Dynamic in Colonial Brazil 87

also Chandler, *The Feitosas and the Sertão dos Inhamuns*, 11.

45. Chandler, *The Feitosas*, 12.
46. Metcalf, "Fathers and Sons," 455–484.
47. Schwartz, *Sugar Plantations*, 272–273.
48. Weber, *Economy and Society*, 1042–1044, 1026–1028.
49. Schwartz, *Sovereignty and Society*, 182.
50. For a relevant typology of orientations, but based on the postcolonial period, see Graham, "Political Power and Landownership in Nineteenth-century Latin America," 129–132.
51. Lang, *Portuguese Brazil*, 146–147.
52. Pang and Seckinger, "The Mandarins."
53. This and the following two paragraphs are based on Schwartz, *Sovereignty and Society*, 280–281, 285–293, 338–343, 173–174, 177–183, 70–71.
54. Dutra, "Centralization vs. Donatorial Privilege"; Lang, *Portuguese Brazil*, 57–70.
55. Schwartz, *Sovereignty and Society*, 188.
56. Ibid., 290; Russell-Wood, "Local Government in Portuguese America," 187–231.
57. See, for example, Rothstein, "The Class Basis of Patron-client Relations," 25–35. For sophisticated empirical studies that compare the consequences of different regional modes of production for interclass relations, see Brustein, "Class Conflict and Class Collaboration in Regional Rebellions, 1500–1700," 445–468; and Brustein, *The Social Origins of Political Regionalism*.
58. This pattern was even more pronounced in the backlands, the frontier, and São Paulo; see Chandler, *The Feitosas*, esp. 33, and Lang, *Portuguese Brazil*, 67–73, on Brazil in general, and 174–184 on Minas Gerais and the stirrings of an independence movement. The relative absence of a royal administration to counter local powers is one factor that may help explain the more autonomous, less "dependent" development of these areas.
59. See, for example, Schwartz, *Sovereignty and Society*, 325–336.
60. Weber, *Economy and Society*, 164–165, 1092, 1094–1095.
61. This and the following two paragraphs are based on Lang, *Portuguese Brazil*, 85–88, 98–102, 105–106; Schwartz, *Sugar Plantations*, 195, 161–162, 181–182.
62. See Weber, *Economy and Society*, 1102–1103.
63. Lang, *Portuguese Brazil*, 116–118, 122.
64. On Portugal's merchants, Lang, *Portuguese Brazil*, 31–33, and cf. Weber, *Economy and Society*, 1094.
65. Lang, *Portuguese Brazil*, 155–174, 184–195; Alden, *Royal Government*, 288ff.
66. On the conditions of patrimonial capitalist moves toward industrialism, cf. Weber, *Economy and Society*, 1098.
67. Lang, *Portuguese Brazil*, 195–200. On the British in Brazilian markets, see also Novais, "A prohibição das manufacturas no Brasil," 145–166.
68. Weber, *Economy and Society*, 1054.
69. I draw on the account by Flory, *Judge and Jury in Imperial Brazil, 1808–1871*, though the interpretive thrust is mine.
70. Prado Júnior, *The Colonial Background of Modern Brazil*.

71. For a historical sketch of the postindependence state as a patrimonial regime, see Roett, *Brazil*, 32–50.

72. Evans, *Dependent Development: The Alliance of Multinational, State, and Local Capital in Brazil.*

73. See, for example, Pang, *Bahia in the First Brazilian Republic*; Uricoechea, *Patrimonial Foundations*. For an alternative analysis on the gains in state autonomy from patrimonial and other interests, see Topik, *The Political Economy of the Brazilian State.*

74. See, for example, Meade, "The Transition to Capitalism in Brazil," 7–26, esp. 24.

75. The extensive/intensive distinction is developed by Mann, *The Sources of Social Power*, vol. 1.

76. Halperin-Donghi, "Dependency Theory," 129.

77. Aronowitz, "A Metatheoretical Critique."

3. From Slavery to Dependence: A Historiographical Perspective

Luís Carlos Soares
Translated by Richard Graham and Hank Phillips

> The means of production may be robbed directly in the form of slaves. But in that case it is necessary that the structure of production in the country to which the slave is abducted admits of slave labour, or (as in South America, etc.) a mode of production appropriate to slave-labour has to be evolved.
> —*Karl Marx*, A Contribution to the Critique of Political Economy

The Slave Mode of Production

Beginning in the early 1960s, discussions in Brazil revolving on Marxist "evolutionary schemes" changed course. Where such scholars had previously accepted at face value five sequential and necessary stages through which all societies had to pass (primitive community, slavery, feudalism, capitalism, and socialism), the publication of works such as those of Ferenc Tokey and Jean Suret-Canale during the late 1950s rekindled discussion of the "Asiatic" mode of production and led to the first radical critiques of that dogmatic scheme.[1] As a consequence of this productive debate, we have come to see that the development of societies is much more complex than had earlier been imagined. They cannot be reduced to a simplistic, linear explanatory scheme presented in "packaged form" and framing the analyses of social structures and modes of production within predetermined, mechanistic, Europe-centered models. Scholars returned to Marx's text[2] and came to understand that the "list" of modes of production could be considerably longer and that developing societies did not necessarily have to evolve through the same stages. Their development in fact has been diverse and has taken a multiplicity of paths. In addition to the concept of "Asiatic" mode of production, the concepts of "ancient," "Germanic," and "Slavic" modes of production were also rehabilitated, and new directions were opened for the investigation of the already accepted modes of production.

The polemic over "precapitalist" social and economic formations and modes of production was enriched in 1965 by Eugene Genovese's *The*

Political Economy of Slavery, in which the author characterized the southern colonial society, and later that of the U.S. South, as "a social system and a civilization with a distinct class structure, political community, economy, ideology, and set of psychological patterns." In writing about the nineteenth-century southern economy, Genovese asserted that as an "economic system" slavery could never be defined as capitalist even though it developed within a world capitalist market and displayed "many ostensibly capitalist features, such as banking, commerce, and credit." He further added that "many precapitalist economic systems have had well-developed commercial relations, but if every commercial society is to be considered capitalist, the word loses all meaning."[3] So began the debate over American slavery as a specific mode of production, different from either the "feudal" or "capitalist" characterizations up to then attributed to it by a large number of scholars.[4]

In 1969 Genovese published a new book, *The World the Slaveholders Made*, in which he sought to elaborate his analysis of the southern slaveholding society in order to further advance the study of "modern slave societies." In this second work studying the rise and decline of American slave regimes, Genovese tried to characterize them as "in a profoundly contradictory entanglement with their European metropolises." According to him, the "more advanced capitalist countries," especially England and Holland, had to resort to "an archaic mode of production" in their expansion into the American continent—even as the capitalist mode of production was gaining ascendancy within their own borders. Viewed that way, American slaveholding would amount to nothing more than a new edition of the slaveholding mode of production that had existed in the ancient Greco-Roman communities: "The rise of the slave systems in the Americas, in contradistinction to the use of slave labor in a peripheral capacity within an essentially wage-labor system, must be understood as the rise of an essentially archaic mode of production."[5]

Genovese went even further. He affirmed that the implantation of slavery in the American continent had a parallel in Eastern Europe, since what took place in both regions was a social consequence of the emerging world market. Thus, the introduction of the "second servitude" in Eastern Europe during the sixteenth century was also a re-creation of an archaic means of production—in this case the feudal system—which survived up until the nineteenth century.[6] What Genovese ended up saying, and it's worth repeating, is that the slave mode of production, as introduced into the Americas, was a repetition of an archaic mode of production. It constituted one of various possible means for the promotion of the primitive and original accumulation of capital, within the very heart of the capitalist expansion taking place from Western Europe beginning in the sixteenth century.

During the early 1970s, however, the publication of Ciro Cardoso's work contributed to the reorientation of the debate among historians of the American slave economy.[7] His work is of fundamental importance, and we can acknowledge that it started the effort to set out new bases for discussions of those modes of production introduced into the American continent beginning in the sixteenth century. Ciro Cardoso laid out as a hypothesis what he called a "theory of colonial modes of production" in the Americas, starting from the premise that "the dynamics of these modes of production are particularly complex, and must be studied while [simultaneously] taking into consideration both internal contradictions and external stimuli, as well as the ways in which these latter became internalized as a function of the former." He uses the term *colonial* not in a political sense, but rather as defining a "relation of structural dependency." We can turn to Ciro Cardoso's own definition: "I designate, then, as 'colonial modes of production,' those modes of production that emerged in the Americas as a function of European colonization, but that were able to survive the American colonies' political independence and continue to exist during the nineteenth century, up until the introduction—which took place in distinct epochs in the various countries—of the capitalist mode of production."[8]

Another point he made is "the subordinate, dependent character of the internal contradictions" within the American colonial societies and "the generally determinant character of the external pressures on the structural changes that took place in those societies." It is incorrect, he quickly adds, to "exaggerate the importance of the colonial aspect and detract from the internal dynamics" of these societies. Nevertheless, American colonial modes of production were at bottom "dependent modes of production." In an effort further to clarify the theoretical foundations of these "dependent modes of production," Ciro Cardoso stated that "dependency—which has as one of its corollaries the transfer of part of the economic surplus to the metropolitan regions—is, due to circumstances surrounding their very genesis and evolution . . . a fact inseparable from the *conception* and structures of said modes of production."[9]

As a preliminary working hypothesis, Cardoso proposed the existence of at least three main "colonial modes of production" upon which American social formations depended: "(1) a mode of production based on the exploitation of Indian labor, established in the central region of pre-Columbian America . . . ; (2) the colonial slave mode of production introduced in regions characterized, on the one hand, by a sparse indigenous population at the time the Europeans arrived, and on the other, by conditions propitious to exporting activities based on plantations producing tropical products or on the exploitation of precious metals (as Minas Gerais's gold, in Brazil) . . . ; [and] (3) in North America 'an autonomous and diversified economy made up of small proprietors,' the only one among the colonial structures able to evolve—even in part during the colonial era—toward industrialization and a capitalism of the 'metropolitan,' not peripheral, type." And he went still further: If these are the three principal modes of production, they coexisted with other, "secondary" modes of production, "established either in the same regions or in subsidiary or bordering areas (such as the cattle-producing regions of colonial Brazil, for example)."[10]

To forestall misunderstandings, the author took great pains to distinguish between the American colonial slave systems and the Greco-Roman classical slave system, emphasizing the differing internal dynamics of these modes of production and also the two distinct moments in history during which each existed:

> The American colonial slave mode of production had the characteristics of a dependent mode of production, since from the outset the corresponding social formations had been dependent, peripheral, and deformed. Here slavery, unlike the slavery of antiquity, was not a consequence of a long evolutionary process; it was the result of a conscious decision made within the context of an effort to create an export economy rapidly.
>
> Unlike the pattern of classical slaveholding, in which slaves were brought a few at a time from societies that had the same level of development as that of the masters—and occasionally an even higher level in certain aspects as was true, for example, of the enslaved Greek preceptors in Rome—in the Americas the practice meant the sudden and brutal impressment, continuously repeated from the sixteenth to the nineteenth centuries, of significant contingents from populations at a lower level of development than those of Europe and belonging to different "races." This led to two important consequences: 1) the brutalizing nature of slavery itself was maximized in the Americas, the slave being no more than an instrument, a human beast of burden meant only for physical toil; 2) conditions attending slavery in the Americas favored the emergence of racial prejudice and posed difficulties for the integration into colonial society of slaves and freedmen. [11]

As we can see, these conclusions of Ciro Cardoso in themselves form a critical response to Eugene Genovese's thesis that slaveholding in the

Americas was no more than a rehash of an ancient mode of production, a "social retrogression" promoted by the expansion of the capitalist economy in Western Europe.[12] The "colonial slave mode of production," argued Cardoso, even though "dependent" on an embryonic world market, possessed an internal functional dynamic of its own, and this was one of the preconditions for the genesis of a capitalist mode of production. This "dependency" persisted through the nineteenth century, when capitalism already predominated in Western Europe. If, on the one hand, the introduction of the colonial slave system in the Americas was determined by the expansion of European commerce, its later development—or lack thereof—cannot be considered a simple unfolding of what was happening in Western Europe, which at that time was undergoing the process of transition from feudalism to capitalism. Capitalism was, in a manner of speaking, taking its first "steps" and was not yet a dominant mode of production. The American "colonial slave mode of production," despite its dependence, constituted a mode of production distinct from capitalism. It is absurd to define it as capitalistic—as do many Marxist historians who, focusing on the circulation of merchandise, mistakenly equate capitalist production with "production for the market" or "profit motive." This practice led Ciro Cardoso to harshly criticize "circulationist" interpretations and their Weberian influences—influences from which Genovese himself was not free.[13]

Following the path opened up by Genovese and Ciro Cardoso, we find Jacob Gorender's *O escravismo colonial* (Colonial slavery), in which the author proposed to deepen the methodological channel opened up by those authors.[14] He sought to "impose a radical inversion of focus: The colonial economy's productive relations need to be studied from the inside out, contrary to past practice which was from the outside in (whether the subject was the patriarchal family or the land tenure system, whether the market or politics)." Gorender recognized that Genovese had "the merit of introducing the legitimate problematic of slave social formation and its specifically corresponding mode of production," but criticized him for having "insufficient zeal in systematically categorizing the economic theory of the U.S. slave system." As for Ciro Cardoso, Gorender regarded him as the worthy elaborator of this problematic, "who, instead of abstracting a unique and undifferentiated 'colonial mode of production,' [properly] limited himself to the concrete elaboration of the colonial slave mode of production"; he went on to state, however, that Ciro Cardoso's study "suffers from the epistemological limitations of models," since he reduced the economic theory of the colonial slave system to "the combination of characteristic qualities."[15]

Gorender's objective was the "systematic categorizing of colonial slave systems." He explained what this effort would entail: "What is needed, in my view, is a general theory of the colonial slave system that would provide a systematic reconstruction of that mode of production as an organic whole, as a unifying set of categories whose necessary interconnections, following as they do from essential determinations, can be formulated into specific laws." Gorender's purpose in this dense work was, then, to elaborate a theory of the "colonial slave mode of production" in accordance with Ciro Cardoso's conceptualization. To this end, he treated it as a historically *new* mode of production, stressing specific laws distinct from the laws pertaining to other modes of production.[16]

The differentiation Gorender established between the colonial slave system and the Greco-Roman slave system was much more detailed than Ciro Cardoso's. Basing himself from the outset on Marx, Gorender conceived of Greco-Roman slavery as "patriarchal slavery," warning that it should not be associated with the "ancient patriarchal family" nor be taken "as a synonym for domestic slavery, for the latter exclusively comprised nonproductive slaves, dedicated to the master's personal service."[17] The author added:

> Patriarchal slavery . . . has a content similar to productive slavery, even though its production is of goods used and consumed within the same economic unit. This, then, is the distinguishing characteristic of patriarchal slavery: the exploitation of the slave aims at the production of an income in kind. A monetary income, when it appears alongside the latter, serves [only] a complementary function.
>
> In its original structure, Greco-Roman slavery possessed this patriarchal character. It developed as a peculiar form of a natural economy, as a combination of self-sustaining productive units. Its production consisted primarily of consumer goods, a portion of which was exchanged for other goods through barter. Albeit strained by the growing infiltration of mercantile relations, this original form survived and prevailed until the demise of the Roman world. . . .
>
> As a consequence, the concrete necessities of the masters imposed a limit to production. So production concentrated on consumer goods that satisfied individual needs and assured the reproduction of the system, all within the sphere of the economic unit. Absolute autarchy, of course, would have been an extremely rare case, which is why the *oikos* partly produced for trade, including profit-making activities. But, as Weber stresses, the ultimate aim of productive activity in the *oikos* was, not

accumulation, but ". . . to meet, in a natural and organized way, the needs of the master."[18]

With the expansion of the Greco-Roman slave economy, there occurred an increase in commercial relations and even an "international commerce in luxury articles," with specialized slaveholding concerns oriented toward mercantile production (the great Sicilian and Carthaginean agricultural undertakings, artisanship, and mining). According to Gorender, however, these constituted "isolated islands in an ocean of natural economy" that continued to predominate. Commercial relations and productive mercantile activities "were unable to dominate the evolution of Roman slavery to the point of converting it from patriarchal to mercantile."[19]

Besides, in fostering the development of productive forces, these activities produced "a disaggregating effect [on] patriarchal slaveholding." And he continued:

> To be a dominant mode of production, mercantile slaveholding would have to be dependent on an external market or, put another way, would have to be colonial. To be sure, Rome made a tributary territory of Sicily, giving rise there to an export-oriented slaveholding agriculture. She could not, however, make herself into an economic colony, nor would she find a market within the Empire's provinces for the product of her slaveholding estates. The contradiction within Ancient Rome's budding mercantile slave system derived from its having been a system within an imperial metropolis and not within a colony. Hence the insoluble historical impasse.[20]

With his characterization of the "colonial slave mode of production," Gorender rejoins Ciro Cardoso and confirms the exclusively economic significance of the "colonial" concept. This is made explicit in that its "main qualities" were: "1) an economy directed primarily toward the external market, relying on that original stimulus for the growth of its productive forces; 2) the exchange of agricultural produce for foreign manufactured products, with consumer goods figuring heavily in the tally of imports; 3) weak or nonexistent control over commercialization in the external market."[21]

Production "dependent" on an external market was the most relevant characteristic—an "essential premise"—for the existence of the "colonial slave mode of production," since the internal market was characterized by its narrowness and lack of elasticity. Given these circumstances, the function of an intermediary between the producing areas and the consuming markets would correspond to commercial capital; the "sphere of circulation" enjoyed some autonomy "in relation to the source of production, without determining the relations of production in each of the extremes."[22] For Gorender, then, the only form in which a mercantile slave system could develop to its fullest potential was "the *colonial* form of slavery, that is, with a mode of production dependent on a metropolitan market."[23]

How should we consider these theses? In reality, what amounted to a political pact between the dominant classes in the Americas on the one hand and the nonautonomous state apparatuses (the colonial governments) on the other was taken by Ciro Cardoso and Gorender to be a central element in the definition of the "pre-capitalist modes of production." We would argue that the colonial situation was indeed a fact, but in itself only a political fact, since the fundamental characteristic of the colonial state apparatus was that direct control over it was not in the hands of the dominant classes there and these governments did not—at least not as a matter of priority—represent these dominant classes' interests. Control was exercised by a corps of functionaries (a bureaucracy) set up by the various monarchies of Europe that were directly linked to the *European* commercial bourgeoisie and nobility. In this sense, the "colonial" regime in the Americas subordinated these apparatuses of state to those existing in Europe.

In our view, the setting up of a colonial government in the Americas had two basic functions: the first was to capture, by means of tributary mechanisms and commercial monopolies, part of the economic surplus otherwise appropriated by the local dominant classes; and the second was to guarantee a commercial monopoly and consequently reserve the American markets for the European bourgeoisie. Due to the intense competition between various factions of the European commercial bourgeoisie, this commercial linkage could only be maintained through rigid political control of the American society and through the commercial expression of that control, that is, the maintenance of a monopoly, the commercial "exclusive."[24] When colonial-type governments began to impede integration of the American productive structure into the world market and the Industrial Revolution's expansion of the capitalist mode of production opened up new economic horizons for the dominant classes in the Americas, these classes did not hesitate to clear the way toward a greater integration into the world economy, breaking the "colonial pact" and forming national states.

To summarize, despite the apparent differences in their definition of the term *colonial*, both Ciro Cardoso, who refers to it as synonymous with "structural dependency," and Gorender, who uses it in the sense of "dependence on foreign markets," define the colonial slave mode of production as a dependent one. The concept of *dependency* is the foundation upon which the entire theory of colonial modes of production in the Americas was erected. Just as the concern of the initial formulators of this problematic—the "dependency sociologists"—was the explanation of how the Latin American economies were incorporated in the world capitalist system, the theory of colonial modes of production represents a variant of the so-called dependency theory. It seeks to explain the insertion of "precapitalist" formations into the world market before capitalism had developed on a world scale.

The major doubt that assails us is related to the possibility of these two historians constructing a theory of modes of production specific to the Americas—especially of slavery—based on such an *ambiguous* concept as dependency. The concept was formulated within a theoretical framework compromised by the national-developmentalist ideology of the Latin American bourgeoisie in a quite recent period in the history of the hemisphere. Clearly, we do not seek to negate the merits of these two historians, who so effectively introduced the need to think in terms of "American continental economies" and their development through the nineteenth century as specific modes of production.

Dependency Theory

"Dependency theory" emerged during the late 1960s as a critical alternative to the "theory of development" and "structural dualism," both of which had figured in the theoretical frame of the UN's Economic Commission on Latin America (ECLA). This formulation's fundamental concern had been to seek an explanation for Latin American underdevelopment. It traced its historical origins, and attempted to outline a strategy of development for Latin American societies that would break out of "external dependence" and aim instead toward "self-sustained" development.[25]

The "dependency theorists," mainly Fernando Henrique Cardoso and Enzo Faletto in their *Dependency and Development in Latin America,* had as their explicit objective the critique of the fundamental concepts used by ECLA (e.g., "traditional societies/modern societies," "development/underdevelopment," "central economies/peripheral economies," etc.). So they constructed their idea of "dependency" as an alternative:

> It would not be sufficient or correct simply to replace the concepts of development and underdevelopment with those of a central and peripheral economy.... The notion of dependency refers to the conditions under which economic and political systems exist and function, revealing the linkage between the two both in the internal

sphere and the external one.... A society may undergo profound changes in its production system without the creation of fully autonomous decision-making centers. Then again, a national society can achieve a certain autonomy of decision without thereby having a production system and an income distribution comparable to those in the central developed countries or even in some peripheral developing countries.[26]

This perspective led to the thesis of "dependent capitalism" as a theoretical alternative to the concept of "underdevelopment." This is quite explicit in a later work by F. H. Cardoso published in O modêlo político brasileiro, where one observes his preoccupation with defining a dependent capitalist economy: "dependent economy . . . means an economy in which the process of accumulation does not come to its full development: the nonexistence or weakness of a capital-goods producing sector has as a consequence that the large-scale reproduction of capital is only carried out within the central economies."[27]

Not far removed from F. H. Cardoso and Enzo Faletto we find Theotonio dos Santos who, in his definition of "dependency" emphasizes the economic aspects of economic relations among "countries":

Dependency is a situation in which a group of countries have their economies conditioned by the development and expansion of another economy. The interdependent relation among two or more economies and between these and world commerce assumes the form of dependency when some countries (the dominant ones) are able to expand and self-propel themselves, while the others (the dependent ones) can only accomplish this as a reaction to such expansion; a reaction which can have a positive or negative effect on their immediate development. In any case, the basic situation leads the dependent countries to an overall situation of backwardness subject to exploitation by the dominant countries.[28]

Finally, another definition of "dependency," which also emphasizes the relations between "countries" or "nations" is provided by Ruy Mauro Maurini:

The Industrial Revolution, which began [the factory system] corresponds chronologically in Latin America to political independence which, won during the early decades of the nineteenth century, gave rise . . . to countries that soon began to gravitate toward England. The movement of goods and, later, of capital, centered on England, while American countries ignored each other. The new countries interacted directly with the English metropolis and, as a function of her needs, began to produce and export raw materials in exchange for consumer manufactures—and debts. . . .

From that moment on, Latin American relations with European capitalist centers took on a definitive structure—the international division of labor—that determined the ultimate course of development within the region. Put differently, it is from that point on that dependency emerges. It is a subordination between formally independent nations, such that relations of production within the subordinate nations are so modified or re-created as to assure the reproduction of dependency. Dependency can thus only lead to further dependency and its end necessarily presupposes suppression of the relations of production that feed it.[29]

From these quotations it is abundantly clear that, despite the critique of ECLA's "developmentalism," the *dependentistas* were unable to proceed beyond the idea of a specifically Latin American social structure, both before and after the global expansion of capitalism. Despite their mention of the internal aspects of class relations, they defined these relations in terms of the opposition between "countries" or "nations," thus shifting the main thrust of the question.[30] For them class relations in the "dependent countries" would reflect the way their internal contradictions were subordinate to the stimuli coming from the "central countries."

In a critique of dependency theory, Francisco Weffort brought out the ambiguity and lack of precision in the concept of "dependency" that arises from its hesitation "between a national approach and a class approach." Weffort explains that within the first "the concept of a Nation operates as a premise for all subsequent analysis of classes and relations of production, [attributing] a national character (real, possible or desirable) to the economy and to class structure." But the second approach alleges that "in the final analysis the real character of the 'national problem'" is determined by the dynamics of relations of production, of class relations. The nation-class ambiguity and *dependentismo's* theoretical imprecision can only be corrected, according to Weffort, by reorienting the problem of Latin American social formations and approaching it from another perspective, which would be single-mindedly framed "in terms of a class perspective, for which the 'national' problem in general (or dependency in general) would not even exist in [an overall] capitalist system, nor would the Nation be conceived of as an explanatory theoretical principle."[31]

Carrying his critique even further, Weffort states that the "dependency" theoreticians still remain on the same ideological side as ECLA, despite all attempts to break free of this school of thought. Since ECLA's notion derives from a national bourgeois reformist ideological camp—as it only points to "the need for national independence, without the concomitant rupture of internal relations of class domination"—the notion of "structural dependency," concomitantly enfolding internal and external relations, was "merely a more radical version within the same ideological camp." That would suggest that dependency theory was only a variant of petit bourgeois radical nationalism.[32]

Historians and the Idea of the "Nation"

To what extent, then, would Weffort's criticisms of *dependentismo* be in some way applicable to the theory of colonial modes of production? The answer to this question requires, first, a demonstration of the interconnections between this theory and the problematic posed by *dependentismo*. In considering the "precapitalist" modes of production introduced in the Americas (especially the "colonial slave mode") as "dependent modes of production," Ciro Cardoso and Jacob Gorender merely transfer backward in time the line of thought formulated for the present, when the capitalist mode of production predominates worldwide. These two authors start from the premise that the social and economic formations based on these dependent modes of production were not autonomous. They were characterized by the transfer of a large parcel of the economic surplus to the European metropoles. Ciro Cardoso and Gorender, moreover, posit a generally determining character of the external stimuli for the changes in "colonial formations." The course of history and any structural changes within these formations would be determined through the internalization of external contradictions and pressures. In no respect was the converse true, as the transformations occurring in the Americas could never have generated conditions that would bring about transformations in the European metropolises. The *colonial* fact would thus be one of the basic premises for the characterization of "colonial modes of production." Finally, accepting "dependency" as a structural given also means that even once those colonial modes of production fell apart, they could only give way to "peripheral," "nonautonomous," or "dependent" capitalist economies, never to true capitalism.

From this perspective, colonial formations in the Americas figure, subjacently, as "nonautonomous nations," that is, as "nations" whose economic and/or political decision-making centers were not yet located within themselves. Early on, then, there must have existed an antagonism between "nations": autonomous nations vs. dependent nations. Even later, when some "dependent nations" achieved political independence and formed sovereign nation-states, their "structural dependency" limited that autonomy and those "nations" could not realize their full potential. In this way, contradictory class relations within the American precapitalist societies, and even in the European ones, are subsumed within the "national problem." This can only serve to deform the analysis of the content of those class relations.

In the face of these problems with the concept of "dependency" and the "theory of colonial modes of production," we prefer a looser theoretical alternative. It seems more coherent, as we attempt to characterize the "precapitalist" modes of production introduced in the Americas during the sixteenth and seventeenth centuries, to begin with the global expansion of European *commerce.* This alternative characterizes these modes of production as modes that emerged and were stimulated by the expansion of commercial capital and that formed part of a structure with a specific historicity, one in which various modes of production coexisted and which embraced various economic and social formations interconnected by commercial capital; commercial capital that was an autonomous element in its relation to the productive sphere and dominant within that global structure.

To this day, when historians speak of the existence of a "Brazilian slaveholding society," the idea they directly convey is that of the existence of a single structure or organization within the area now known as the "Brazilian territory," a structure whose development began in the sixteenth and continued through the latter half of the nineteenth century. Even some Marxist historians, encouraged by the emergence of the debate concerning social and economic formations and insensitive to the company they keep, have actually come out and stated, following the same convention, that there had existed a *"Brazilian* economic and social slaveholding formation." During the course of our research, however, we came to doubt such affirmations and began to advance the hypothesis that from the sixteenth through the nineteenth centuries there existed not one, but many precapitalist economic and social formations within the area that is today Brazil.

This questioning came about when we undertook a critical review of the "theory of dependency" and its intimate relationship with the idea of a "nation." From there, we verified that one commonly finds, not only among *dependentistas* but also among social scientists of various schools, a characterization of "society" or of the "social structure" associated with the idea of a "nation." We will not return here to dependency theory, nor develop an exhaustive polemic with the various "nationalist" schools of thought. We seek only to demonstrate critically the relation between the national project and the views of some historians.

As is well known, from the earliest times the various areas comprising the Portuguese overseas empire in the Americas were generically referred to as "Brazil."[33] In 1621 there was a politico-administrative division of these areas that resulted in the creation of two large states: the state of Brazil and the state of Maranhão and Grão Pará. This division lasted until the middle of the eighteenth century when the Marquis of Pombal, in an effort to erect a centralized political and administrative machine, reunited the Portuguese dominions in the Americas and created a single state of Brazil, with the city of Rio de Janeiro, at least nominally, as its capital. After the arrival of the Portuguese Royal Family and the transfer of the seat of the monarchy to Rio de Janeiro in 1808, this city became the focus of attention for the other captaincies. As the capital of an enormously extended empire, the city was the nerve center of all political goings on, and government leaders were now in much closer contact with the slaveholding planters. Rio was also the nucleus of an entire bureaucratic-tributary machine, set up by King João VI in order to guarantee financial survival.

The country's political independence did not immediately alter that scheme. With the Brazilian Empire founded and her constitution decreed in 1824, the goal of a unitary state with provinces subordinated to the center in Rio de Janeiro triumphed. Effective control of the state by the slaveholding plantation owners only came about, however, after Pedro I's abdication in 1831 and the beginning of the regency. At that time, sectors of the classes dominant in various regions of the empire began to engage in debate and, concomitantly, in political action with the intent of establishing a new model for the government. The proposed alternatives were varied, but all can be ranged between the two poles of centralization and decentralization.

Unitarism, as consecrated by the Constitution of 1824, was effectively challenged by the partial victory of decentralizing proposals in 1834 and with the promulgation that year of the Additional Act to the constitution, which established a relative provincial autonomy, and, on another level, by the outbreak of political revolts in some provinces. But soon the faction headed by Bernardo Pereira de Vasconcelos halted the most exalted and decentralizing liberals' "revolutionary carriage," took over the leadership of the state, and imposed "the *return* of authority and order." In other words, the project for a unitary and centralized monarchy was taken up again in 1837 with the toppling of Diogo Antonio Feijó and subsequent ascension of Pedro de Araújo Lima.

With the victory of the "return," a concrete problem presented itself: How to make the planters and other social sectors from various parts of the empire, each with distinct immediate interests, come to feel united in an authentic political, economic, and cultural community, the interests and aspirations of which might supersede the more immediate ones of those particular groups and localities? The "nation" was the ideological project of slaveholding plantation owners in response to that question. They had reached a consensus (even if it was imposed by a fraction of their class on all the others and on "civil society," that is, on the free). It amounted in effect to the social construction of a "Brazilian nation."

In building the "Brazilian nation" there was much room for history and historians. The Brazilian Institute of History and Geography began its work just as unitarism and centralization emerged victorious. Beginning with its foundation in 1838, it produced a history that emphasized national unity, retelling the glories of the "nation" and defending "great men" who were intimately linked to the slaveholding planters.[34] Canon Januário da Cunha Barbosa, the secretary of the institute, clearly spelled out in a report written in 1840 the directives for historical investigation that historians and members of the society ought to observe when writing national history (historia patria):

> History, in making present the experience of past centuries, ministers us counsel, both true and disinterested, that illuminates the course to be taken, the shoals to be avoided, and the secure port into which a wise maneuver can felicitously guide the ship of State. . . .
> Politics and civilization in general demand that we apply ourselves to saving from the voracity of the centuries those facts that have conducted us to our present state and that will, in future times, serve as a point of comparison for our progress after our establishment as an independent nation. Witness of the times, light of truth, she abounds with elements necessary to our civilization and the prosperity of the State; as instructress of life, she offers examples of heroic deeds for those who prize the honor of serving the fatherland.[35]

In the dedication to the Emperor Pedro II of his *História geral do Brasil* (General history of Brazil) (1854), Francisco Adolpho de Varnhagen, one of the major historians of the last century, reaffirmed in the most succinct way those same directives for historical investigation: "Your Imperial Majesty, my Lord, has recognized the importance and sanctioned the study of the Nation's History, as much to contribute to her greater splendor among foreigners as to provide facts useful in the administration of the State; and also to strengthen the ties of national unity, enliven and exalt patriotism, ennoble public spirit, and augment faith in our future."[36] In this vision of a "history" that teleologically focused all past events as though they were inherent in the process of forming the "nation," forming "Brazil," the dominant slaveholding class projected onto the pasts of the various areas of the country all that it desired and needed in order to consolidate its power over the recently created state.

Even later, during the twentieth-century formation and development of Brazilian capitalism, the bourgeoisie's intellectuals accepted that historical view. Indeed they began with it, only touching up some of its points in an effort to furnish an ideological framework to serve as a basis for the attainment of "national integration," that is to say, of a national market.[37] As for the so-called popular movements in our past, even to this day official historiography views them with a certain disdain, as movements that placed the "harmony of the nation" at risk.

Some historians, however, influenced by national-developmentalism—especially José Honório Rodrigues—criticized official history for being written on behalf of "often alienated elites and antinational interests." He spoke of the need to incorporate "the people" as central to History, which is to say, that there be developed "a new vision of History that recognizes popular virtues and accomplishments and incorporates them into the very bloodstream of national politics."[38] But in fact, the national-developmentalist historians appealed to a romanticized vision of "the people," in the style of that liberalism that accompanied the great European bourgeois revolutions.[39] Still lurking behind the need to emplace "the people" as central to history is the idea of the full construction of the "nation," of the construction of a true community of individuals with common traditions, aspirations, and interests.[40]

Even Caio Prado Júnior himself, the first to break with historiography founded on the ideological premises of the dominant classes and to try, instead, to think out our history from the Marxist perspective of social classes with their antagonisms and contradictions, even he was unable to escape from the specter of the "nation." His books are heavily marked by a nationalist vision.[41] The perspective of a "national" history, which "the ruling classes and their historians elaborated with the center as starting point,"[42] has exerted a strong influence on Marxist historians generally, and certainly on Ciro Cardoso and Gorender.[43] For they have affirmed the existence of a "Brazilian" social formation or a "Brazilian" economic and social slaveholding system. According to them, to be sure, there existed economically and socially distinct regions within the "Brazilian" territorial area; but, implicitly if not explicitly, they maintained that by virtue of these regions' identity with each other as determined by their future politico-administrative unity, they always constituted a unique and united economic and social formation.

Advances in historical knowledge now require a break with this centralist and nationalist perspective, and the proposal of a new investigative hypothesis related to the possibility of existence—within the present Brazilian territory—of diverse precapitalist economic and social formations, even if most of them were slaveholding. During the latter half of the nineteenth century, these various precapitalist social and economic formations disintegrated. Throughout the world, in fact, precapitalist formations crumbled before the domineering ascent of the capitalist mode of production. At this time, capitalism's global domination was not established merely—as had long been true—at the level of circulation of merchandise. It was also and mainly (beginning in the 1860s) established at the production level with the export of productive capital from the originally capitalist economies to the previously noncapitalist ones. The latter, already undergoing transformations resulting from their ever-increasing integration into the world market, were unable to withstand the impact of this penetration of capital and readily succumbed to the onset of capitalist development.

The second half of the nineteenth century was marked by an authentic "world revolution" because the expanding capitalist mode of production managed—not without some conflicts and contradictions—to integrate all corners of the world; transforming the old precapitalist formations into new areas of capitalist accumulation. With the beginning of the current century, the accelerated reproduction of capital was already taking place, integrating all of these areas into a single worldwide productive and mercantile circuit. From this profound "revolution," caused by capitalism's expansion, there at last emerged, in truth, a worldwide social and economic formation in which the areas corresponding to the old defunct precapitalist formations were transformed into mere parts or regions of a larger whole.

During the first half of the current century, the interconnections among the various areas of "Brazil" were finally accomplished. Providing the principal impetus for these ties, one finds the process of industrialization in the Southeast. The need for capitalism to expand throughout the "Brazilian" territory overcame the other productive regions and made them into sources of manual labor and raw materials for southeastern industries and into consumer markets for products of these same industries. An economic homogenization took place, and the different regions practically disappeared to make room for a great arena of capital accumulation and reproduction that enveloped the entire "national" area, having as its dynamic nucleus the South-Central producing areas, mainly around São Paulo.[44] This transformation of the "Brazilian" territorial area into a sole, large, capitalist region, albeit featuring localized zones of dissimilar productive activities, can evidently only be understood within the general context of the transformations originating in the global expansion of monopolistic capitalism and, consequently, in the tendencies toward centralization of capital, monopolization of the economy, and the homogenizing of the economic areas that accompany these trends. Toward the middle of the twentieth century, yes, the inhabitants of the most different and far-flung parts of the country had come to think of themselves as, and therefore actually to be, "Brazilians." Thus, due to the development of imperialist capitalism, "national integration" was finally brought about.

Notes

1. Quoted in Garaudy, Chestneaux, Godelier, et al., *O modo de produção asiático*, 14. Also see Galissot, Parain, Vilar, et al., *Sur le feodalisme*; we used the Portuguese translation: *Sobre o feudalismo*.

2. Especially *Die formen der Kapitalistichen Produktion Verhergehen* (1858), published in Spanish under the title "Formas que preceden a la producción capitalista," in Marx, *Elementos fundamentales*, vol. 1.

3. Genovese, *The Political Economy of Slavery*, 3, 19.

4. Among those historians and social scientists who classified slaveholding in Brazil as "feudal," using the existence of landed estates as a presupposition for such a characterization, we may highlight: Duarte, *A ordem privada e a organização política nacional*, 2d ed., chap. 2; Guimarães, *Quatro séculos de latifúndio*, chap. 2. Among those who characterize slaveholding as "capitalist," not just in Brazil, but in all slaveholding areas in the Americas, and who use market-oriented and profitmaking production as a paradigm of capitalism, we may note especially: Prado Júnior, *A revolução brasileira*, chap. 3; Beiguelman, *Formação política do Brasil*, vol. 1, chap. 1; F. H. Cardoso, *Capitalismo e escravidão no Brasil meridional*, chap. 4; Franco, *Homens livres na ordem escravocrata*, "Introdução"; and Frank, *Capitalismo y subdesarrollo en América Latina*, chap. 3. There are some who defend the thesis of "commercial capitalism" or, more precisely, of the insertion of a slave economy into the larger scope of "commercial capitalism," and among these the most important is Novais, *Estrutura e dinâmica do Antigo Sistema Colonial*. Also see Mauro, *Nova história e Novo Mundo*, and Williams, *Capitalismo e escravidão*, chaps. 3 and 8. In addition to these "feudal" and "capitalist" characterizations, we might also note the position of Rangel who, using a dualist approach, defines the "Brazilian economy" from the beginning of colonization to the end of slavery as "a slave-capitalist duality" and as "internally slaveholding and externally capitalist," *Dualidade básica da economia brasileira*; and that of Sodré who, although recognizing the existence of differences between "colonial slaveholding" as introduced in Brazil and elsewhere in the Americas and the slaveholding of antiquity, follows the "Stalinist scheme" of the five necessary and successive stages in the evolution of societies and ends up defining both the American and ancient systems as "slave modes of production," *Formação histórica do Brasil*.

5. Genovese, *The World the Slaveholders Made*, v, 21, 22.

6. Ibid., 22–23.

7. C. F. S. Cardoso, "Severo Martínez Peláez y el carácter del régimen colonial," "Sobre los modos de producción coloniales de América," and "El modo de producción esclavista colonial en América," all in Assadourian et al., *Modos de producción en América Latina*, 83–109, 135–159, 193–242.

8. Ibid., 142.
9. Ibid., 142, 152; emphasis in original.
10. Ibid., 153, 154.
11. Ibid., 224
12. Genovese, *World the Slaveholders Made*, 23.
13. Besides Ciro Cardoso's works already cited, see also his "Los modos de producción coloniales," 87–105. There is also Ciafardini's "Capital, comercio y capitalismo," 111–134.
14. Gorender, *O escravismo colonial*.
15. Ibid., 21, 22
16. Ibid., 22, 55.
17. Ibid., 166.
18. Ibid., 166–167.
19. Ibid., 168.
20. Ibid.
21. Ibid., 170.
22. Ibid., 169, 171.
23. Ibid., 169, 170, 171; emphasis in original.
24. On this commercial monopoly or "exclusive," compare Novais, *Estrutura e dinâmica do Antigo Sistema Colonial*, with Gorender, *Escravismo colonial*, 489–523.
25. Two syntheses of ECLA's concepts are: Furtado, *Subdesenvolvimento e estagnação na América Latina*; and Sunkel, *O marco histórico do processo desenvolvimento-subdesenvolvimento*.
26. F. H. Cardoso and E. Faletto, *Dependência e desenvolvimento na América Latina*, 27 [Ed.: I relied on but modified the English edition: *Dependency and Development in Latin America*, 18–19.]
27. F. H. Cardoso, *O modêlo político brasileiro*, 43. It is worth mentioning that in another chapter F. H. Cardoso affirms that it is impossible to think of a "theory of dependency," for the idea of "elevating the notion of dependency to the category of an all-encompassing theory is *nonsense*." Thus, instead of speaking of a "theory of dependency," it would be much more correct, he says, to speak of the possibility of "concrete analyses of dependent situations," which would be nothing more than "the political expression, at the periphery, of the capitalist mode of production when it expands at the international level" (ibid., 128). Despite this caveat, it has become commonplace to refer to it as a theory.
28. Santos, "La crisis de la teoria del desarrollo," 180.
29. Maurini, *Tres ensayos sobre América Latina*, 99–100.
30. The *dependentistas* have never clearly explained what they mean by "nation" or "country," but implicit in their use of these two terms are the traditional meanings attributed to them by social scientists: Thus, "nation" can be viewed as a "human community inhabiting the same territory and [having shared] a common origin, customs and interests for a long time, and, usually, having a single language.... A nation has a past in common of which its members have some knowledge. The whole of it is simultaneously political, economic and normally cultural, expressing itself, or trying to express itself, through common institutions," Birou, *Dicionário das ciências sociais*, 271. Each "country" can be seen as "a unit, one of the zones into which the world is divided with attention

to diverse geographical elements. The correspondence between physical and anthropo-geographic elements forms a unit with its own characteristics," Maull, *Geografía Política*, 104.

31. Weffort, *O populismo na política brasileira*, 170, 173.

32. Ibid., 170.

33. Useful starting points for the general political history of Brazil are the following: Prado Júnior, *Formação do Brasil contemporâneo* (The colonial background of modern Brazil); Prado Júnior, *Evolução política do Brasil e outros estudos*; Faoro, *Os donos do poder*, I, 169–275, 313–337; Costa, "Introdução ao estudo da emanicipação política do Brasil"; Costa, *Da monarquia à república: momentos decisivos* (The Brazilian empire: myths and realities), chap. 3; Beiguelman, *Formação política do Brasil*, vol. 1.

34. Rodrigues, *A pesquisa histórica no Brasil*, 37–40.

35. Barbosa, "Relatorio do Secretario Perpetuo," 573.

36. Varnhagen, *Historia geral do Brazil*, I, i–ii.

37. See, on the role of intellectuals, Gramsci, *A formação dos intelectuais*. Among the Brazilian bourgeoisie's "organic intellectuals"—to use Gramsci's phrase for those organically tied to a dominant class—who produced explanations of Brazil's "national" economy and historic formation, those who stand out are: Simonsen, *História econômica do Brasil (1500–1820)*; Simonsen, *Evolução industrial do Brasil e outros estudos*; Furtado, *Formação econômica do Brasil*.

38. Rodrigues, *História e historiadores do Brasil*, 10.

39. On the meanings of "people" and "nation" in the nineteenth century, see Hobsbawm, *A era das revoluções, 1789–1848*, chap. 7.

40. Here is how Rodrigues handles the idea of "nation": The "national interest is what serves the interests of the common people and the federal union, in whose name the Nation was constituted." Or again: "Nations, as political societies, are dominated by vital interests which give rise to emotional reactions and rational convictions and are responsible for a historical unity in the behavior of peoples and their leaders. All [nations] have permanent aspirations, which are the product of history," Rodrigues, *Aspirações nacionais*, 9–10, 79. [Ed.: I have here drawn on the English version: *The Brazilians: Their Character and Aspirations*, xx, 69].

41. For example, in his description of the colonial era: "In the colonial past, the foundations of nationhood were laid: a semi-desert territory was settled, and a way of life was established there that differed as much from the life of the indigenous population as it did, on a smaller scale, from that of the Portuguese who had undertaken the task of settlement. Something new had been created in the sphere of human achievement," Prado Júnior, *Formação do Brasil contemporâneo*, 6 [Ed.: I have drawn on the English translation by Suzette Machado, cited in note 33.]

42. Linhares, *História do abastecimento*, 155.

43. See Ciro F. S. Cardoso and H. P. Brignoli, *Los métodos de la historia*, 84, 85; Gorender, *O escravismo colonial*, 170, 291; Levy, *História da Bolsa de Valores do Rio de Janeiro*, 36–38.

44. See Oliveira, *Elegia para uma re(li)gião*.

Bibliography

Alden, Dauril. *Royal Government in Brazil, with Special Reference to the Administration of the Marquis of Lavradio, Viceroy, 1769–1779.* Berkeley: University of California Press, 1968.
Almeida, Candido Mendes de, ed. *Codigo Philippino, ou Ordenações e leis do reino de Portugal* (1st ed. 1603). Rio de Janeiro, 1870.
Anderson, Perry. *Passages from Antiquity to Feudalism.* London: NLB, 1974.
Antonio, Robert J., and Ronald M. Glassman, eds. *A Marx-Weber Dialogue.* Lawrence: University Press of Kansas, 1985.
Aronowitz, Stanley. "A Metatheoretical Critique of Immanuel Wallerstein's *The Modern World System.*" *Theory and Society* 19 (1981): 503–520.
Assadourian, Carlos Sempat, et al. *Modos de producción en América Latina,* special issue of *Cuadernos de Pasado y Presente,* no. 40. 3d ed. Buenos Aires: Siglo XXI, 1975.
Atkinson, R. F. *Knowledge and Explanation in History: An Introduction to the Philosophy of History.* Ithaca, N.Y.: Cornell University Press, 1978.
Azevedo, João Lúcio de. *Épocas de Portugal econômico: esboços de história.* 2d ed. Lisbon, 1947.
Bach, Robert L. "On the Holism of the World-systems Perspective." In *Processes of the World-System,* edited by Terence K. Hopkins and Immanuel Wallerstein. Beverly Hills, Calif.: Sage, 1980, 289–310.
Barbosa, Januário da Cunha. "Relatorio do Secretario Perpetuo." *Revista trimestral de historia e geographia, ou jornal do Instituto Historico e Geographico Brasileiro,* tome II, 1840 (2d ed. 1858).
Beiguelman, Paula. *Formação política do Brasil.* São Paulo: Pioneira, 1967.
Bendix, Reinhard. *Force, Fate, and Freedom: On Historical Sociology.* Berkeley: University of California Press, 1984.
Bendix, Reinhard, and Guenther Roth. *Scholarship and Partisanship: Essays on Max Weber.* Berkeley: University of California Press, 1971.
Birou, A. *Dicionário das ciências sociais.* Translated by Alexandre Gaspar. Lisbon: Dom Quixote, 1973.
Booth, David. "Marxism and Development Sociology: Interpreting the Impasse." *World Development* 13 (1985): 761–787.
[Brandão, Ambrósio Fernandes?] *Diálogos das grandezas do Brasil* (1618). Edited by J. A. Gonsalves de Mello. Recife, 1966.

Braudel, Fernand. *La Méditerranée et le monde méditerranéen à l'époque de Philippe II*. Paris, 1949.
Brenner, Robert. "The Origins of Capitalist Development: A Critique of Neo-Smithian Marxism." *New Left Review* 104 (July/August 1977): 25–92.
Brustein, William. "Class Conflict and Class Collaboration in Regional Rebellions, 1500–1700." *Theory and Society* 14 (1985): 445–68.
———. *The Social Origins of Political Regionalism: France, 1849–1981*. Berkeley: University of California Press, 1988.
Buss, Andreas E. *Max Weber in Asian Studies*. Leiden: E. J. Brill, 1985.
Cardoso, Ciro Flamarion Santana. "Los modos de producción coloniales: estado de la cuestión y perspectiva teórica," *Estudios Sociales Centroamericanos* 4:10 (Jan.–Apr. 1975), 87–105.
———. "Severo Martínez Peláez y el carácter del régimen colonial." In *Modos de producción en América Latina*, a special issue of *Cuadernos de Pasado y Presente*, no. 40 (Cordoba, 1973).
———. "Sobre los modos de producción coloniales de América." In *Modos de producción en América Latina*, a special issue of *Cuadernos de Pasado y Presente*, no. 40 (Cordoba, 1973).
Cardoso, Ciro Flamarion Santana, and Hector P. Brignoli. *Los métodos de la historia*. Barcelona: Editorial Crítica-Grijalba, 1976.
Cardoso, Fernando Henrique. *Capitalismo e escravidão no Brasil meridional: o negro na sociedade escravocrata do Rio Grande do Sul*. Rio de Janeiro: Paz e Terra, 1977.
———. "The Consumption of Dependency Theory in the United States." *Latin American Research Review* 12, no. 3 (1977): 7–24.
———. *O modêlo político brasileiro*. São Paulo: DIFEL, 1972.
Cardoso, Fernando Henrique, and Enzo Faletto. *Dependência e desenvolvimento na América Latina: ensaio de interpretação sociológica*. Rio de Janeiro: Zahar, 1970.
———. *Dependency and Development in Latin America*. Translated by Marjory Mattingly Urquidi. Berkeley: University of California Press, 1979.
Cartas jesuíticas. Rio de Janeiro: Edição da Academia Brasileira, 1931.
Chandler, Billy Jaynes. *The Feitosas and the Sertão dos Inhamuns: The History of a Family and a Community in Northeast Brazil, 1700–1930*. Gainesville: University Presses of Florida, 1972.
Chase-Dunn, Christopher. "Commentary on Robert L. Bach, 'On the Holism of a World-systems Perspective.'" In *Processes of the World System*, edited by Terence K. Hopkins and Immanuel Wallerstein. Beverly Hills, Calif.: Sage, 1980, 312–313.
Chirot, Daniel, and Thomas D. Hall. "World-system Theory." *Annual Review of Sociology* 8 (1982): 81–106.
Ciafardini, Horácio. "Capital, comercio y capitalismo: a propósito del llamado 'capitalismo comercial'," in Carlos Sempat Assadourian et al., *Modos de producción en América Latina*, special issue of *Cuadernos de Pasado y Presente*, no. 40. 3d ed. Buenos Aires: Siglo XXI, 1975, 111–134.
Clark, Victor. *History of Manufactures in the United States*. New York, 1949.
Clough, Shepard B., and Charles W. Cole. *Economic History of Europe*. Boston, 1952.

Bibliography

Cohen, Sande. *Historical Culture.* Berkeley: University of California Press, 1986.
Collins, Randall. *Weberian Sociological Theory.* New York: Cambridge University Press, 1986.
Costa, Emília Viotti da. *The Brazilian Empire: Myths and Realities.* Chicago: University of Chicago Press, 1985.
———. *Da monarquia à república: momentos decisivos.* São Paulo: Grijalbo, 1977.
———. "Introdução ao estudo da emanicipação política do Brasil." In *Brasil em perspectiva.* Edited by Carlos Guilherme Mota. São Paulo: DIFEL, 1971.
Dias, Manuel Nunes. *O capitalismo monárquico português (1415–1549).* Coimbra, 1963.
Dobb, Maurice. *Studies in the Development of Capitalism.* London, 1954.
Duarte, Nestor. *A ordem privada e a organização política nacional.* 2d ed. São Paulo: Pioneira and Editora Nacional, 1966.
Dutra, Francis A. "Centralization vs. Donatorial Privilege: Pernambuco, 1602–1630." In *Colonial Roots of Modern Brazil,* edited by Dauril Alden. Berkeley: University of California Press, 1973, 19–60.
Eisenstadt, S. N. *The Political Systems of Empires.* New York: Free Press, 1963.
Eisenstadt, S. N., and René Lemarchand, eds. *Political Clientelism, Patronage, and Development.* Beverly Hills, Calif.: Sage, 1981.
Evans, Peter. *Dependent Development: The Alliance of Multinational, State, and Local Capital in Brazil.* Princeton, N.J.: Princeton University Press, 1979.
Faoro, Raymundo. *Os donos do poder: formação do patronato político brasileiro.* Porto Alegre and São Paulo: Editora Globo and EDUSP, 1975 (1st ed. 1958).
Fernandes, Florestan. *Sociedade de classes e subdesenvolvimento.* 2d ed. Rio de Janeiro, 1972.
Flory, Thomas. *Judge and Jury in Imperial Brazil, 1808–1871.* Austin: University of Texas Press, 1981.
Franco, Maria Sylvia de Carvalho. *Homens livres na ordem escravocrata.* São Paulo: Atica, 1974.
Frank, Andre Gunder. *Capitalism and Underdevelopment in Latin America: Historical Studies of Chile and Brazil.* New York: Monthly Review Press, 1967.
———. *Capitalismo y subdesarrollo en América Latina.* Translated by Elpidio Pacios. Buenos Aires: Siglo XXI, 1973.
———. "Development and Underdevelopment in the New World: Smith and Marx vs. the Weberians." *Theory and Society* 2 (1975): 431–466.
Furtado, Celso. *The Economic Growth of Brazil.* Berkeley: University of California Press, 1963.
———. *Formação econômica do Brasil.* [? ed.] São Paulo: Editora Nacional, 1971 (1st ed. Rio de Janeiro, 1959).
———. *Subdesenvolvimento e estagnação na América Latina.* Rio de Janeiro: Civilização Brasileira, 1968.
Galissot, René, Charles Parain, Pierre Vilar, et al. *Sobre o feudalismo.* Translated by Maria de Fatima Martins Pereira. Lisbon: Editorial Estampa-CERM, 1973.

———. *Sur le féodalisme*. Paris: Editions Sociales, 1971.
Garaudy, Roger, Jean Chestneaux, Maurice Godelier, et al. *O modo de produção asiático*. Lisbon: CERM-Seara Nova, 1974.
Garst, Daniel. "Wallerstein and His Critics." *Theory and Society* 14 (1985): 469–495.
Genovese, Eugene D. *The Political Economy of Slavery: Studies in the Economy & Society of the Slave South*. New York: Pantheon, 1965.
———. *The World the Slaveholders Made: Two Essays in Interpretation*. New York: Vintage Books, 1971.
Godinho, Victor Magalhães, ed. *Documentos sôbre a expansão portuguesa*. Lisbon, 1943.
Gorender, Jacob. *O escravismo colonial*. São Paulo: Atica, 1978.
Goulart, Maurício. *A escravidão africana no Brasil*. 2d ed. São Paulo, 1950.
Graham, Richard. "Political Power and Landownership in Nineteenth-century Latin America." In *New Approaches to Latin American History*, edited by Richard Graham and Peter H. Smith. Austin: University of Texas Press, 1974, 112–136.
———. "State and Society in Brazil, 1822–1930." *Latin American Research Review* 22 (1987): 223–236.
Graham, Richard, and Peter H. Smith. "Introduction." In *New Approaches to Latin American History*, edited by Richard Graham and Peter H. Smith. Austin: University of Texas Press, 1974, ix–xiv.
Gramsci, Antonio. *A formação dos intelectuais*. Translated by Serafim Ferreira. Lisbon: Venda Nova-Amadora, 1972.
Guimarães, Alberto Passos. *Quatro séculos de latifúndio*. São Paulo: Fulgor, 1964.
Hall, John R. "Where History and Sociology Meet: Modes of Discourse and Analytic Strategies." Paper presented at the annual meetings of the American Sociological Association, Atlanta, Georgia, 1988.
———. "World-system Holism and Colonial Brazilian Agriculture." *Latin American Research Review* 19, no. 2 (1984): 43–69.
Halperin-Donghi, Tulio. "'Dependency Theory' and Latin American Historiography." *Latin American Research Review* 17, no. 1 (1982): 115–130.
Hauser, Henri, and A. Renaudet. *Les débuts de l'âge moderne*. 4th ed. Paris, 1956.
Heckscher, Eli F. *La época mercantilista*. Spanish transl. Mexico, 1943.
Himmelfarb, Gertrude. *The New History and the Old*. New York: Cambridge University Press, 1987.
Hobsbawm, Eric J. *A era das revoluções, 1789–1848*. Translated by Maria Tereza L. Teixeira and Marcos Penchel. Rio de Janeiro: Paz e Terra, 1977.
———. "The Revival of Narrative: Some Comments." *Past and Present* no. 86 (February 1980): 3–8.
Holanda, Sérgio Buarque de, and Olga Pantaleão. "Francêses, holandêses, e inglêses no Brasil quinhentista." In *História geral da civilização Brasileira*, edited by Sérgio Buarque de Holanda. São Paulo, 1960, tomo 1, vol. 1, 147–175.
Howe, Gary N., and Alan M. Sica. "Political Economy, Imperialism, and the Problem of World System Theory." *Current Perspectives in Social Theory*

1 (1980): 235-286.
Johnson, H. B., Jr. "The Donatory Captaincy in Perspective: Portuguese Backgrounds to the Settlement of Brazil." *Hispanic American Historical Review* 52 (1972): 203-214.
Keynes, John Maynard. *Teoria geral do emprego, do juro e do dinheiro*. Portuguese transl. Rio de Janeiro, 1964.
Krieger, Leonard. *Ranke: The Meaning of History*. Chicago: University of Chicago Press, 1977.
Laclau, Ernesto. "Feudalism and Capitalism in Latin America." *New Left Review* 67 (May-June, 1971), 19-38.
Lang, James. *Portuguese Brazil: The King's Plantation*. New York: Academic Press, 1979.
Lash, Scott, and Sam Whimster, eds. *Max Weber, Rationality, and Modernity*. London: Allen & Unwin, 1987.
Leroy-Beaulieu, Paul. *De la colonisation chez les peuples modernes*. Paris, 1874 (5th ed. 1902).
Levy, Maria Bárbara. *História da bôlsa de valôres do Rio de Janeiro*. Rio de Janeiro: IBMEC, 1977.
Lewin, Linda. *Politics and Parentela in Paraíba: A Case Study of Family-Based Oligarchy in Brazil's Old Republic*. Princeton, N.J.: Princeton University Press, 1987.
———. "Some Historical Implications of Kinship Organization for Family-based Politics in the Brazilian Northeast." *Comparative Studies in Society and History* 21 (1979): 262-292.
Linhares, Maria Yedda L. *História do abastecimento: uma problemática em questão (1530-1918)*. Brasília: BINAGRI, 1979.
Love, Joseph L. "An Approach to Regionalism." In *New Approaches to Latin American History*, edited by Richard Graham and Peter H. Smith. Austin: University of Texas Press, 1974, 137-155.
Luton, Harry. "The Satellite/Metropolis Model: A Critique." *Theory and Society* 3 (1976): 573-581.
Mandel, Ernest. *Long Waves of Capitalist Development*. New York: Cambridge University Press, 1980.
Mann, Michael. *The Sources of Social Power, Vol. I: A History of Power from the Beginning to A.D. 1760*. New York: Cambridge University Press, 1986.
Marchant, Alexander. "Feudal and Capitalistic Elements in the Portuguese Settlement of Brazil." *Hispanic American Historical Review* 22 (1942): 493-512.
Marx, Karl. *A Contribution to the Critique of Political Economy*. Moscow: Progress Publishers, 1970.
———. *El Capital: crítica de la economía política*. Spanish transl. Mexico: Editora Fondo de Cultura, 1946.
———. *Elementos fundamentales para la crítica de la economía política (borrador) 1857-1858*. Translated by Pedro Scaron. Buenos Aires: Siglo XXI, 1973.
Marx, Karl, and Friedrich Engels. "Sobre el colonialismo." Spanish transl. In *Cuadernos de Pasado e Presente*, no. 37 (Cordoba, 1973).
Maull, Otto. *Geografía política*. Translated by Ismael Antich. Barcelona:

Omega, 1960.
Maurini, Ruy Mauro. *Tres ensayos sobre América Latina.* Barcelona: Cuadernos Anagrama, 1973.
Mauro, Frédéric. *Le Portugal et l'Atlantique au XVIIe Siècle, 1570–1670: étude economique.* Paris: SEVPEN, 1960.
———. *Nova história e Novo Mundo.* Portuguese transl. São Paulo: Perspectiva, 1969.
———. "Recent Works on the Political Economy of Brazil in the Portuguese Empire." *Latin American Research Review* 19, no. 1 (1984): 87–105.
Meade, Teresa. "The Transition to Capitalism in Brazil: Notes on a Third Road." *Latin American Perspectives* 5, no. 3 (Summer 1978): 7–26.
Metcalf, Alida C. "Fathers and Sons: The Politics of Inheritance in a Colonial Brazilian Township." *Hispanic American Historical Review* 66 (1986): 455–484.
Mörner, Magnus, Julia Fawaz de Viñuela, and John D. French. "Comparative Approaches to Latin American History." *Latin American Research Review* 17, no. 3 (1982): 55–89.
Mukherjee, Ramkrishna. "Commentary on Robert L. Bach, 'On the Holism of a World-systems Perspective'." In *Processes of the World-System*, edited by Terence K. Hopkins and Immanuel Wallerstein. Beverly Hills, Calif.: Sage, 1980, 314–316
Novais, Fernando A. "Colonização e sistema colonial: discussão de conceitos e perspectiva histórica." In IV Simpósio Nacional dos Professores Universitários de História, *Anais.* São Paulo, 1969.
———. *Estrutura e dinâmica do Antigo Sistema Colonial (séculos XVI–XVIII).* São Paulo: Brasiliense, 1974.
———. "Portugal e Brasil na crise do Antigo Sistema Colonial (1777–1808)." PhD diss., University of São Paulo, 1972.
———. *Portugal e Brasil na crise do antigo sistema colonial (1777–1808).* São Paulo: HUCITEC, 1979.
———. "A prohibição das manufacturas no Brasil e a política econômica portuguêsa do fim do século XVIII." *Revista de História* (São Paulo), no. 67 (1967): 145–166.
Oliveira, Francisco de. *Elegia para uma re(li)gião: Sudene, Nordeste, planejamento e conflitos de classes.* Rio de Janeiro: Paz e Terra, 1977.
Pang, Eul-Soo. *Bahia in the First Brazilian Republic: Coronelismo and Oligarchies, 1889–1934.* Gainesville: University Presses of Florida, 1979.
Pang, Eul-Soo, and Ron L. Seckinger. "The Mandarins of Imperial Brazil." *Comparative Studies in Society and History* 14, no. 2 (1972): 215–244.
Prado Júnior, Caio. *The Colonial Background of Modern Brazil.* Translated by Suzette Macedo. Berkeley: University of California Press, 1967.
———. *Evolução política do Brasil e outros estudos.* São Paulo: Brasiliense, 1972.
———. *Formação do Brasil contemporâneo (Época colonial).* São Paulo: Brasiliense, 1963 (4th ed. 1953).
———. *A revolução brasileira.* São Paulo: Brasiliense, 1966.
Rangel, Ignacio. *Dualidade básica da economia brasileira.* Rio de Janeiro: ISEB, 1957.

Bibliography

Rodrigues, José Honório. *Aspirações nacionais: interpretação histórico-política*. São Paulo: Fulgor, 1965.
———. *The Brazilians: Their Character and Aspirations*. Translated by Ralph Edward Dimmick. Austin: University of Texas Press, 1967.
———. *História e historiadores do Brasil*. São Paulo: Fulgor, 1965.
———. *A pesquisa histórica no Brasil*. São Paulo: Cia. Editora Nacional, 1969.
Roett, Riordan. *Brazil: Politics in a Patrimonial Society*. New York: Praeger, 1975.
Romano, Ruggiero. "American Feudalism." *Hispanic American Historical Review* 64, no. 1 (1984): 121–134.
Roscher, W., and R. Jannasch. *Kolonien, Kolonialpolitik und Auswanderung*. 3d ed. Leipzig, 1885 (1st ed. 1848).
Roth, Guenther. "History and Sociology in the Work of Max Weber." *British Journal of Sociology* 27 (1976): 306–318.
———. "Personal Rulership, Patrimonialism, and Empire-building." In Reinhard Bendix and Guenther Roth, *Scholarship and Partisanship: Essays on Max Weber*. Berkeley: University of California Press, 1971, 156–169.
———. "Rationalization in Max Weber's Developmental History." In *Max Weber, Rationality, and Modernity*, edited by Scott Lash and Sam Whimster. London: Allen & Unwin, 1987, 75–91.
———. "Sociological Typology and Historical Explanation." In Reinhard Bendix and Guenther Roth, *Scholarship and Partisanship: Essays on Max Weber*. Berkeley: University of California Press, 1971, 109–128.
Rothstein, Frances. "The Class Basis of Patron-client Relations." *Latin American Perspectives* 6, no. 2 (1979): 25–35.
Russell-Wood, A. J. R. "Local Government in Portuguese America: A Study in Cultural Divergence." *Comparative Studies in Society and History* 16, no. 2 (1974): 187–231.
Santarém, Visconde de. *Memórias e alguns documentos para a história e teoria das Cortes Gerais*. Lisbon, 1924.
———. *Quadro elementar das relações politicas e diplomaticas de Portugal*. Paris, 1842.
Santiago, Theo A., ed. *América colonial: ensaios*. Rio de Janeiro, 1975.
Santos, Theotonio dos. "La crisis de la teoria del desarrollo y las relaciones de depêndencia en América Latina." In *La dependencia político-económica de la América Latina*, edited by Aldo Ferrer et al. México: Siglo XXI, 1973.
Schluchter, Wolfgang. *Rationalism, Religion, and Domination: A Weberian Perspective*. Berkeley: University of California Press, 1989.
Schwartz, Stuart B. *Sovereignty and Society in Colonial Brazil: The High Court of Bahia and Its Judges, 1609–1751*. Berkeley: University of California Press, 1973.
———."State and Society in Colonial Spanish America: An Opportunity for Prosopography." In *New Approaches to Latin American History*, edited by Richard Graham and Peter H. Smith. Austin: University of Texas Press, 1974, 3–35.
———. *Sugar Plantations in the Formation of Brazilian Society: Bahia, 1550–1835*. New York: Cambridge University Press, 1985.
Sée, Henri. *As origens do capitalismo moderno*. Portuguese transl. Rio de

Janeiro, 1959.
Simonsen, Roberto C. *Evolução industrial do Brasil e outros estudos*. São Paulo: Editora Nacional, 1973.
———. *História econômica do Brasil (1500–1820)*. [? ed.] São Paulo: Editora Nacional, 1969.
Skocpol, Theda, ed. *Vision and Method in Historical Sociology*. New York: Cambridge University Press, 1984.
———. "Wallerstein's World Capitalist System: A Theoretical and Historical Critique." *American Journal of Sociology* 82 (1977): 1075–1090.
Smith, Adam. *An Inquiry into the Nature and Causes of the Wealth of Nations*. (1st ed. 1776). Edited by E. Cannan. New York: Modern Library, n.d.
Sodré, Nelson Werneck. *Formação histórica do Brasil*. São Paulo: Brasiliense, 1971.
Sorre, Maximilien. *Les migrations des peuples*. Paris, 1955.
Stark, W. *Historia de la economía en su relación con el desarollo social*. Spanish transl. Mexico, 1961.
Stein, Stanley J., and Barbara H. Stein. *The Colonial Heritage of Latin America: Essays on Economic Dependence in Perspective*. New York: Oxford University Press, 1970.
Stern, Steve J. "Feudalism, Capitalism, and the World-System in the Perspective of Latin America and the Caribbean." *American Historical Review* 93:4 (Oct. 1988), 829–872.
Stone, Lawrence. "The Revival of Narrative: Reflections on a New Old History." *Past and Present* no. 85 (November 1978): 3–24.
Sunkel, Oswaldo. *O marco histórico do processo desenvolvimento-subdesenvolvimento*. São Paulo and Rio de Janeiro: DIFEL and Forum, 1975.
Sweezy, Paul, et al. *The Transition from Feudalism to Capitalism*. Introduction by Rodney Hilton. London: Verso, 1976.
Taylor, Kit Sims. *Sugar and the Underdevelopment of Northeastern Brazil, 1500–1970*. Gainesville: University Presses of Florida, 1978.
Tilly, Charles. *Big Structures, Large Processes, Huge Comparisons*. New York: Russell Sage, 1984.
Topik, Steven. *The Political Economy of the Brazilian State, 1889–1930*. Austin: University of Texas Press, 1987.
Trimberger, Ellen Kay. "World Systems Analysis: The Problem of Unequal Development." *Theory and Society* 8 (1979): 127–137.
Turner, Bryan S. *For Weber*. London: Routledge and Kegan Paul, 1981.
———. "The Structuralist Critique of Max Weber's Sociology." *British Journal of Sociology* 28 (1977): 1–16.
Uricoechea, Fernando. *The Patrimonial Foundations of the Brazilian Bureaucratic State*. Berkeley: University of California Press, 1980.
Varnhagen, Francisco Adolfo de. *Historia geral do Brazil*. Rio de Janeiro: Laemmert, n.d.
Verlinden, Charles. *The Beginnings of Modern Colonization*. Ithaca, N.Y.: Cornell University Press, 1970.
Veyne, Paul. *Writing History*. Middletown, Conn.: Wesleyan University Press, 1984.

Vicens-Vives, J., ed. *História social y económica de España y America*. Barcelona, 1957.
Wallerstein, Immanuel. *The Capitalist World-Economy: Essays*. New York: Cambridge University Press, 1979.
———. *The Modern World-System: Capitalist Agriculture and the Origins of the European World-Economy in the Sixteenth Century*. New York: Academic Press, 1974.
———. "The Rise and Future Demise of the World Capitalist System: Concepts for Comparative Analysis." *Comparative Studies in Society and History* 16:4 (Sept. 1974), 387–415.
Walton, John. "Small Gains for Big Theories: Recent Work on Development." *Latin American Research Review* 22, no. 2 (1987): 192–201.
Weber, Max. *The Agrarian Sociology of Ancient Civilizations*. London: NLB, 1976.
———. *Economy and Society*. Berkeley: University of California Press, 1978.
———. *General Economic History*. New Brunswick, N.J.: Transaction, 1981.
———. *Wirtschaftsgeschichte*. 3d ed. Berlin, 1958.
Weeks, John, and Elizabeth Dore. "International Exchange and the Causes of Backwardness." *Latin American Perspectives* 6, no. 3 (1979): 62–87.
Weffort, Francisco. *O populismo na política brasileira*. Rio de Janeiro: Paz e Terra, 1978.
White, Hayden. "The Question of Narrative." *History and Theory* 32 (1984): 1–33.
Williams, Eric. *Capitalism and Slavery*. 2d ed. New York, 1961.
———. *Capitalismo e escravidão*. Translated by Carlos Neyfeld. Rio de Janeiro: Editora Americana, 1975.
Zirker, Daniel. "Brazilian Development: Alternative Approaches to an Increasingly Complex Field." *Latin American Research Review* 18, no. 2 (1983): 135–149.
Zurara, Gomes Eanes de. *Crônica dos feitos da Guiné*. Edited by A. J. Dias. Lisbon, 1949.

Index

Absolutist systems, 14–15, 16, 17, 22
Africa, 22, 24, 25
African slaves, 2, 29, 40, 44–45. *See also* Slavery
Agriculture, 68–72. *See also names of specific crops*
Amsterdam Company, 24
Anderson, Perry, 65
Antilles, 30, 36, 37, 40, 53
Aronowitz, Stanley, 83
Atlantic islands, 24–25, 27, 35, 36, 44
Azevedo, Lúcio de, 44
Azores, 27

Barbosa, Januário da Cunha, 103
Bilateral oligopoly, 33
Bourgeoisie, 17, 19, 24–25, 32, 39, 42, 46, 47, 64, 96, 97, 100, 104
Braudel, Fernand, 17, 58
Brazil: administration of, 65–68, 74–75; administration of justice in, 67–68, 74–75; agriculture in, 68–72; capitalism in, 104; captaincies in, 2, 25–26, 64, 65–68, 68, 74, 75; colonization by Portugal, 1–2, 21, 27–28, 36, 102; families and patronage in, 73–77; family wealth in, 72–73; independence of, 2, 80–82, 102; landowners in, 68–72, 74–76, 79–80, 102, 103; legacy of patrimonialism in postindependence Brazil, 80–82; mercantilism and capitalism in, 77–80; as nation, 102–104; patrimonialism in, 5–6, 61–80; relocation of Portuguese court to, 79, 102; slavery in, 6–7, 44–45, 102–103, 106n.4; social classes in, 61, 69–70, 76; wealth of, 54n.9
Brazil Company, 78
Brazilian Institute of History and Geography, 103
Britain. *See* England
British East Indies Company, 30
Bullion, 13, 17
Bureaucratic class, 74, 75, 76

Capital: accumulation of, 16, 19, 20, 32, 36, 51, 58, 90, 105; commercial, 96, 101; exports of, 105; among merchant classes, 32; primitive accumulation of, 21, 32, 35, 38, 39, 40, 43, 45, 46, 50, 90; reproduction of, 105
Capitalism: in Brazil, 104; and colonialism, 2, 3; commercial, 17, 35, 39, 49, 51, 106n.4; dependent, 7, 98; historical development of, 3–4, 14, 18, 19–20, 41, 89, 93; industrial, 18, 20, 35, 50, 51–52, 61; mercantile, 14, 16, 18, 20, 22, 32, 35, 38, 43, 51; modes of production of, 90, 91–92, 93, 100, 105; monarchic, 22–23; patrimonial, 65, 79; and patrimonialism, 6, 69, 77–80, 82; and Protestant ethic, 60;

requirements for, 71; shaping of, 83; and slavery, 2, 3, 48
Cardoso, Ciro F. S., 7, 91, 92, 93, 96, 100, 104
Cardoso, Fernando Henrique, 7, 8, 97, 98
Castillo, Céspedes del, 29
Cattle industry, 46, 68
Charles III, 28
Child, Josiah, 30, 51
Cities. See Metropolis
Class relations, 49–50, 61, 69–70, 76, 99–100, 101. See also Bourgeoisie; Bureaucratic class; Merchants; Seigneurial class
Coelho, Duarte, 1, 26
Coffee, 1, 2, 68, 69
Colbert, Jean Baptiste, 31
Colonial economies, 17–18, 25–26, 35–39, 40–41, 43–44, 46–50
Colonial Pact, 12, 53
Colonialism: advantages of, 34–35; in Atlantic islands, 24–25, 27, 35, 36; and bourgeoisie, 24–25; and capitalism, 2, 3–4; and colonial economies, 17–18, 25–26, 35–39, 40–41, 43–44, 46–50; and commercial expansion, 18–19; competition among European nations, 29–30; crisis of mercantilist colonialism, 45–53; definition of Old Regime, 17; of England, 4, 30–31, 36, 52–53; exclusivist mechanism of, 32–34; exploitation colonies, 20–21, 37; and exports, 4–5; and foreign concessions, 27, 28; of France, 29, 31; of Italy, 66; legislation concerning, 11–12; and mercantilism, 11–21, 30; and metropolitan exclusive, 12, 14, 21–35, 37–38, 39, 47, 49, 53–54n.3; modes of production of, 91–92, 93, 95–97; monopolistic approach to, 24–26, 27, 28; in North America, 20–21, 37–38, 51; and ports in the Americas, 28–29; of Portugal, 1–2, 21, 27–28, 35–36, 54n.17, 102; and prohibitionist policy, 50–51; settlement colonies, 20–21, 30, 37, 38, 40, 51; and slavery, 39–45, 46–48, 89–97; and social categories, 49–50; of Spain, 28–29, 36; as system, 11–21. See also Brazil; Dependency theory; Mercantilism
Commerce. See Capitalism; Exports; Mercantilism; Trade routes; Trading companies
Commercial capitalism, 17, 35, 39, 49, 51, 106n.4
Contraband. See Smuggling
Corruption, 74, 76
Cotton, 31, 35
Council of the Indies, 27
Country, definition of, 107–108n.30. See also Nation
Cromwell, Oliver, 30–31

Dependency theory: and idea of a "nation," 99–100, 101; approach of, 8, 97–98; and class relations, 99–100; definitions of dependency, 97–99; dependency sociologists 97; dependent capitalism, 7, 98; dependent nations, 100–101; and Latin American history, 59; limitations of, 107n.27; and patrimonialism, 79, 81–82; and slavery, 6–7; structural dependency, 96–97, 100; and world-system theory, 58
Development, theory of, 97
Diamonds, 1, 67
Dobb, Maurice, 3, 15
Dos Santos, Theotonio, 7

East Indies Company, 24, 30, 31
ECLA. See Economic Commission on Latin America
Economic Commission on Latin America (ECLA), 8, 97, 99–100
Economic systems. See Capitalism; Colonialism; Mercantilism

Index

Endogamous marriages, 73
England: Act of 1660, 31; Acts of Trade, 12; capitalism in, 90; colonialism of, 4, 20, 30–31, 36, 52–53; economic activities of, 15, 24, 78; and Latin American countries, 98–99; as nation-state, 16; Navigation Acts, 4, 30; North American colonies of, 4, 30, 37–38; relationship to Portugal, 27, 79; Staple Act, 31; Treaty of Utrecht, 29
Europe. *See names of specific countries*
Exogamous marriages, 73
Exploitation colonies, 20–21, 37
Exports, 1–2, 4–5, 13, 30–31, 37, 38, 43–44, 46–47, 70–71, 77, 79, 105

Faletto, Enzo, 97, 98
Families, 72–77
Faoro, Raymundo, 62
Feijó, Diogo Antonio, 102
Feudalism, 3, 5, 14, 15, 16, 18, 19, 39, 40, 41, 42, 51, 63, 64, 65–67, 77, 81, 89, 90, 93
Flanders, 15, 22, 24
France, 15, 16, 29, 30, 31, 36, 40, 66
Frank, Andre Gunder, 8, 57, 65
Freyre, Gilberto, 2
Fur trade, 35, 48
Furtado, Celso, 26, 48

Gee, Joshua, 30
General Commerce Company, 27
Genovese, Eugene D., 7, 89–90, 92–93
Gold, 1, 13, 36, 67, 92
Gorender, Jacob, 7, 93–94, 96, 100, 104
Gouveia, Diogo de, 36
Great Britain. *See* England
Greco-Roman classical slave system, 92, 94–95
Grotius, 23
Guinea Company, 31

Heckscher, Eli F., 15
Henry, Prince, 22, 25, 35, 54n.14

Holland. *See* Netherlands, 23

Import substitution, 79
India, 22, 25, 27, 66
Indians, 2, 91
Indies, 23, 29
Indigo, 31, 35
Industrial capitalism, 18, 20, 35, 50, 51–52, 61
Industrial Revolution, 4, 11, 14, 19, 39, 42, 50, 51–52, 96, 98
Italy, 15, 22, 23, 24, 25, 30, 66

Jesuits, 44–45
João III, 66
João VI, 102
John III, 36

Keynes, John Maynard, 54n.6
Kinship, 72–73

Labor. *See* Slavery; Wage labor
Laclau, Ernesto, 8
Landowners, 68–72, 74–76, 79–80, 102, 103
Lenin, Vladimir Ilyich, 58
Leroy-Beaulieu, Paul, 20
Lima, Araújo, 102
Long, James, 67

Madeira Islands, 1, 24, 27, 36, 66
Manuel I, 24
Marriage, 72–73, 76
Marx, Karl, 19, 41, 42, 58, 62, 64, 69, 89, 94
Marxism, 43, 93, 101, 104
Maurini, Ruy Mauro, 7, 98
Mauro, Frédéric, 26
Mercantile capitalism, 14, 16, 18, 20, 22, 32, 35, 38, 43, 51
Mercantilism: and colonial economies, 38–39, 40–41; and colonization, 4, 11–21; crisis of mercantilist colonialism, 45–53; factors counteracting, 33–34; and metropolitan "exclusive," 21–35; and overseas competition, 52–53; and patrimonial-

ism, 77–80; policy of, 13–14, 16–17, 19–20, 50–51; theory of, 12–13, 31
Merchants, 32, 33, 34, 35, 61, 67, 70, 72, 74, 77–78, 79
"Metalist" idea, 13
Metals, 18, 35, 92
Metcalf, Alida, 73
Metropolis: and colonialism, 21, 26–27; development of, 12, 14; metropolitan exclusive and colonial system, 21–35, 37–38, 39, 47, 49, 53–54n.3; and primitive accumulation of capital, 21
Middle Ages. See Feudalism
Mining, 36, 95. See also names of minerals and metals
Modernization theory, 62
Monarchic capitalism, 22–23
Monopoly, definition of, 33. See also Mercantilism
Monopsony, 33
Morazé, Charles, 17
Mun, Thomas, 13, 30

Napoleon, 79
Nation: definition of, 107n.30, 108nn.40–41; formation of, 16; idea of, 99–106
Navigation Acts, 4, 30
Netherlands, 16, 23–24, 26, 27, 29–30, 36, 77–78, 90
Nóbrega, Manoel da, 44
Noronha, Fernando de, 25
North American colonies, 4, 30, 31, 37–38, 51, 90

Old Colonial System. See Colonialism
Oligopsony-oligopoly, 33, 71–72
Orient, 15, 23, 24, 31, 33, 54n.17, 66

Patrimonial capitalism, 65, 79
Patrimonial state, 63, 73–74
Patrimonialism: and administration of colonial Brazil, 65–68, 73–74; and agricultural production, 68–72; in colonial Brazil, 64–80; definition of, 5–6; durable organization of, 63–64; and family wealth, 72–77; legacy of, 80–82; and mercantilism and capitalism, 77–80; and patronage, 73–77; power involved in, 62–63; and slavery, 69–70; Weberian approach to, 61–64; and world-system theory, 58, 83
Patron-client relationships, 69–70, 76
Patronage, 73–77, 81
Pedro I, 102
Pedro II, 2, 103
Philippine Ordinances, 27
Phillip II, 23
Piracy, 28, 31, 78
Pombal, Marquis de, 6, 79, 102
Portugal: annexation by Spain, 23; colonialism of, 1–2, 21, 27–28, 35–36, 54n.17, 102; control of the *asiento*, 29; economy of, 22, 64; expansion during 1400s, 22; exports from colonies of, 30–31; independence from Spain, 27; kings of, 5–6, 33, 64, 65, 67, 77; as nation-state, 16; Overseas Council of, 27; and patrimonial system, 5–6, 65, 79–80; relocation of court to Brazil, 79, 102; royal monopoly in, 22–23; uniqueness of, 64
Postlethwayt, Malachy, 12, 30
Prado Júnior, Caio, 18, 81, 104
Prebendal feudalism, 65–67
Proletarianization, 15
Protectionist policy, 13, 17
Protestant ethic, 60

Richelieu, Armand Jean du Plessis, 31
Rodrigues, José Honório, 104
Roett, Riordan, 62
Roman classical slave system. See Greco-Roman classical slave system
Roth, Guenther, 60

Index

Royal and Supreme Council of the Indies, 28

Santos, Theotonio dos, 98
Schwartz, Stuart, 59, 73, 75
Seigneurial class, 46–47, 48, 50, 65, 70, 72–73
Senegal Company, 31
Settlement colonies, 20–21, 30, 37, 38, 40, 51
Silver, 13, 36
Slave trade, 2, 29, 31, 39–40, 44, 66, 79
Slavery: abolition of, 69; African slaves, 2, 29, 40, 44–45; in Brazil, 6–7, 44–45, 102–103, 106n.4; and capitalism, 2, 3; and colonial economy, 39–45, 46–48, 89–97; and dependency theory, 6–7; and exploitation colonies, 21; Greco-Roman classical slave system, 92, 94–95; of Indians, 2; Marxist analysis of, 43; as mode of production, 32, 89–97; opposition to, 39, 44–45; and patrimonialism, 69–70; in U.S. South, 90, 91, 92–93
Smith, Adam, 13, 41
Smuggling, 27, 29, 34, 52–53
Social classes. *See* Bourgeoisie; Bureaucratic class; Class relations; Merchants; Seigneurial class
Socialism, 89
Sorre, Maximilien, 11
Spain, 15, 16, 23, 24, 26–31, 33, 36
Staple Act, 31
Stark, W., 17
Stein, Barbara, 8
Stein, Stanley, 8
Structural dualism, 97
Sugar, 1, 24, 25–26, 31, 35, 36, 49, 66, 68–71, 77
Suret-Canale, Jean, 89
Sweezy, Paul, 3

Tariffs, 13, 17

Technology, 47
Timber, 1, 2, 25, 31, 35, 37, 48
Tobacco, 31, 35, 37, 68, 69
Tokey, Ferenc, 89
Trade. *See* Exports; Mercantilism; Merchants
Trade routes, 15, 18, 22, 29
Trading companies, 23–24, 29–30, 31, 33, 37, 78
Treaty of 1810, 79
Treaty of Utrecht, 29

Underdevelopment, 98
Union of Utrecht, 23
United Nations Economic Commission on Latin America, 8, 97, 99–100
United States: independence of, 53; slavery in, 90, 91, 92–93
Universal history, 57, 82
Urban areas. *See* Metropolis
Uricoechea, Fernando, 62

Vargas, Getúlio, 2
Varnhagen, Francisco Adolpho de, 103
Vasconcelos, Bernardo Pereira de, 102

Wage labor, 39, 41, 42, 90
Wallerstein, Immanuel, 8, 58, 62, 64
Wealth. *See* Capital
Weber, Max, 5, 20, 57–64, 65, 66, 69, 71, 73–74, 75, 77, 80, 83, 93, 94–95
Weffort, Francisco, 7, 99–100
West Indies, 27, 70
West Indies Company, 30, 31
Williams, Eric, 43, 44
Wood. *See* Timber
World-system theory: concept of, 8; critical reaction to, 58–59; limitations of, 5, 58–59, 62, 82–83; Weberian approach to, 57–61

Zurara, Gomes Eanes de, 44